Astrobiology of Earth

Astrobiology of Earth

The emergence, evolution, and future of life on a planet in turmoil

JOSEPH GALE

OXFORD
UNIVERSITY PRESS

OXFORD

UNIVERSITY PRESS

Great Clarendon Street, Oxford OX2 6DP

Oxford University Press is a department of the University of Oxford.
It furthers the University's objective of excellence in research, scholarship,
and education by publishing worldwide in

Oxford New York

Auckland Cape Town Dar es Salaam Hong Kong Karachi
Kuala Lumpur Madrid Melbourne Mexico City Nairobi
New Delhi Shanghai Taipei Toronto

With offices in

Argentina Austria Brazil Chile Czech Republic France Greece
Guatemala Hungary Italy Japan Poland Portugal Singapore
South Korea Switzerland Thailand Turkey Ukraine Vietnam

Oxford is a registered trade mark of Oxford University Press
in the UK and in certain other countries

Published in the United States
by Oxford University Press Inc., New York

© Joseph Gale 2009

The moral rights of the authors have been asserted
Database right Oxford University Press (maker)

First published 2009

British Library Cataloguing in Publication Data

Data available

Library of Congress Cataloging in Publication Data

Data available

Typeset by Newgen Imaging Systems (P) Ltd., Chennai, India
Printed in Great Britain
on acid-free paper by
the MPG Books Group

ISBN 978–0–19–920580–6 (Hbk.)   978–0–19–920581–3 (Pbk.)

1 3 5 7 9 10 8 6 4 2

Contents

vi Contents

Introduction

Astrobiology is a 'new' science that presumes to study life in the universe. Life does exist in the universe. The evidence for this can be seen on our own planet, Earth. Moreover the enormous size of the known cosmos makes it highly unlikely that we are unique and that life does not exist on a planet circling our own or some other star. Our galaxy, the Milky Way, contains some 100 billion[1] stars and is only one of some 100–200 billion galaxies. However, to date there is no proof that life exists anywhere other than on Earth. Given this single known occurrence it is not possible to calculate the statistical odds of finding extraterrestrial life. Today most astrobiologists believe that some form of life exists in many other places in the cosmos, even in our galaxy. However, the possibility of the existence of intelligent life is considered to be much, much smaller.

On Earth only an extraordinary, fortuitous combination of environmental conditions allowed life to emerge, survive, and evolve from single cells to intelligent multicellular animals and to continue to exist today. How this came about is a mystery that is slowly being unravelled. Astrobiologists study the conditions that allowed life to appear on Earth and search for similar conditions on planets in the Solar System and in other star systems. As a working hypothesis and starting point for this search they use the initial conceit that extraterrestrial life will be similar to our own, and would therefore depend on similar environmental conditions; a preconception long abandoned by science fiction writers!

Astrobiology was described above as a new science, but with qualification of the word 'new'. This is because the question of the possible existence of extraterrestrial life has in fact been discussed ever since humanity wrote books, and probably long before. *Astrobiology of Earth* is limited to and focused on the events—astronomical, geological, and biological—which allowed life to appear and evolve on our planet, finally producing us. In the words of Pierre Charron in the 16th century: 'The true science and study of man is man'. Forgiving Charron his now politically incorrect 'man', he made his point, which is the point of this book.

The common thesis of ecology is that the environment of the currently human-friendly biosphere[2] depends on a delicate and unique combination of numerous factors, mainly of Earth origin. Astrobiology shows how these life-enabling factors are first determined by extraterrestrial parameters. These include Earth's position and motions

---

[1] One billion is defined as a thousand million (10^9). The SI prefix for 10^9 is G (see table of SI prefixes in Symbols and Abbreviations below). The British billion was formerly defined as a million million (10^{12}): this is now obsolete in scientific and most general usage.

[2] The biosphere is the thin layer on planet Earth from about 3 km below the surface to a few kilometres above where life exists.

within the Solar System, the position of the Solar System within the Milky Way, and the position of our galaxy in the cosmos. To this must be added the radiation environment of the Earth, which is determined mainly by the Sun but also by extra-Solar System radiation and, to a small extent, by the Earth itself.

In addition to Earth's position in the Solar System its life favourable conditions are, of course, also determined by planet Earth's own chemical and physical characteristics. For example, one of the most important of these is its density and gravity that have allowed the retention of an atmosphere, which is neither too thin nor too massive for the needs of the biosphere.

Many other cosmic bodies, apart from the Sun, affect the biosphere. They include supernovae and, nearer to us, the Solar System planets, comets, asteroids, and our Moon. An appreciation of these factors is consequently essential to our understanding of the special physical, chemical, and biological parameters of the Earth's biosphere. These extraterrestrial and geological factors are all highly dynamic. They have changed frequently over the lifetime of the Earth, strongly affecting life—sometimes catastrophically.

As a result of these changes, the biosphere does not present a constant and benign environment. At best it is in a state of quasi-equilibrium, one that humanity is today capable of perturbing. The environment could rapidly degrade into one in which humans could no longer prosper. One extreme example is that, following a world-wide nuclear war, humans would probably disappear while cockroaches could thrive. We would not be the first Earth life form to so affect the biosphere. Billions of years ago plant photosynthesis degraded the environment (from the point of view of plants). Even today, after billions of years of evolution, most plant species are partly starved and partly poisoned by today's plant-determined atmosphere. This same atmospheric composition is, however, very clement to the plants' competitors, such as insects and animals, including humans. Only an understanding of all the factors whose complex interactions produced the present biosphere will enable us to appreciate how human activity may upset its delicate equilibrium, and suggest ways to avoid an apocalyptic outcome. Obviously we wish to continue to survive in good health, feed the growing world population, and allow all the peoples of the world to attain the standard of living enjoyed by the top 20% today. It is currently unclear how this will come about when, in about 50 years from now, the world population reaches the predicted 9–10 billion (up from 6 billion at the turn of the millennium). These goals are not unattainable. Solutions based on science are for the most part known. The present world problems of overpopulation, pollution, starvation, and disease are mainly technical and political.

It has often been suggested, especially by science fiction enthusiasts, that humanity could solve the problems of the pollution and depletion of Earth's resources by the exploration and settlement of other planets and star systems. The final chapter of this book addresses this proposition.

With our own interests in mind, we must first understand what produced and currently maintains the biosphere in its human-favourable conformation, and then choose our policies wisely. To quote once again the almost proverbial and intentionally provocative question:

Never-mind the universe; is there intelligent life on Earth?

A small confession—with a purpose

When proposing a new book to a publisher the would-be author is often asked about her or his motivation for writing on the particular subject. I wrote saying that during many years of teaching environmental physiology I came to recognize the effects on and importance to plants and animals of numerous factors of the larger environment from geology to astronomy.

Moreover, a parallel career in military and civilian aviation had led me to carry out research at the US National Aeronautics and Space Administration (NASA), within their space biology, controlled environment life support systems (CELSS) programme. From there the distance to astrobiology was short. All this is perfectly true, but it is not the whole story and was not perhaps the main incentive.

It began many years ago. As a child of the Second World War, in London, my parents sent me to a countryside summer camp in Cambridgeshire. The official reason was to participate in scouting activities. The real intention was to get away, if only for a few months, from the continuing German blitz. The latter is an essential part of the story. England in those days was 'blacked out' to give the Luftwaffe navigational problems (not that that worked). So I spent a midsummer far from the disturbing lights of town or village. 'Disturbing' yes, for seeing the heavens. Looking up into the night sky on the rare cloudless days, one could see a sight to which few town girls or boys, like me, had ever been exposed. It seemed as though the entire universe was in our hands. Lying on the grass we could make out the stars and constellations, sometimes 'shooting stars' (meteors) and even the Milky Way. It was a humbling and transfixing experience.

A few years later I found myself in the southern Israeli Negev desert. On a moonless night in the desert 'one can see the glories of the Lord'. There (and there) and then I decided to become an astronomer (no, not a priest, kadi, or rabbi). However, after a short time I realized that one needed far more mathematical ability than I could ever muster for the physics that is an astronomer's daily bread.

The above very personal admission does not mean that the less mathematically gifted cannot appreciate the wonders of the heavens, including our own planet, and their possible effect on our lives. The evidence of our eyes and the guidance of astronomers, physicists, geologists, and biologists can bring this about. So, for a start, go out on a clear moonless night, far from towns, villages, and freeways. If possible go to the top of a mountain, where the air is thin and clear, or on a boat a few miles from shore, or anywhere where the lights of 'civilization' are not disturbing. There you will see the real starting point for an intellectual odyssey: wonder at the size of the universe, a yearning to know how we came to be here, how this great universe affects and effects our lives, and whether we are alone or perhaps have neighbours somewhere in our galaxy looking towards us with the same questions. Finally, observing the heavens and studying Earth and ourselves makes us aware of the fortuity, smallness, insignificance, provisionality, ephemerality, fragility, and hence extreme value of life on our planet.

A note on prerequisites

Astrobiology of Earth is a multidisciplinary field. This is reflected in the students. Their majors may be from general biology, astronomy, earth sciences, or ecology, with some coming from other disciplines, with little science background. To cater for all, the material is presented here very basically, at a level suitable for undergraduates. This means that astronomy majors may skip sections on the size and content of the universe, biology students must forgive being explained the nature of RNA and DNA, and so on. All that is required is a good high-school science level, preferably one or two years of university-level science, in any department, and an open mind. The basic didactic concept followed here is that students should be able to see the forest and not just the trees, as is so common in today's for the most part highly reductionist science courses. Students wishing to pursue these subjects for higher degrees are referred to specialist departments such as astronomy, biology, or earth sciences.

Each of the chapters is intended as a guide to students preparing class/term papers. They are almost independent. Consequently, in the interest of clarity, there is some overlap which, in a learning environment, is rarely harmful. The rather extensive list of references and resources should be useful in the preparation of term papers and for readers wanting to dig further.

Acknowledgements

Astrobiology is very eclectic. A specialist in just one subject (and very few are not these days) has to be a little courageous to undertake to lead such a workshop. I could not have undertaken this course and textbook without the help of many experts outside of my own limited field of expertise. Moreover, here, the adage of a Talmud teacher of 1700 years ago, 'from all my students I gathered knowledge', is particularly apt. Many thanks to all of you.

The late Robert. D. MacElroy, my mentor at NASA, deserves particular mention and thanks for introducing me to the biology of space research and encouraging this writing project. It is sad that he is not around to see it accomplished.

I owe much to my colleagues, in different branches of science, who I burdened with editing the chapters nearest to their own specialities. Special thanks go to Yigal Erel, Raphael Falk, Gideon Fleminger, Joseph Herschberg, Yaakov Lorch, Nir Shaviv, and Joseph Sperling, each of whom reviewed one or more chapters. I am especially grateful to Alfred M. Mayer, my one time teacher and long-time colleague, a true polymath who read every word and made many corrections and insightful comments. However, and as usual, any remaining mistakes are entirely my responsibility; I would be only too happy to hear from you, the reader.

Doron Eisenstadt, my TA for this course over the last few years, was invaluable in preparing the manuscript for the publishers.

Last, but perhaps first, thanks to my wife Rena Gale, a medical researcher and physician, my severest critic, who commented on each chapter and patiently suffered through the usual tribulations of living with someone writing a textbook.

Symbols and abbreviations

AU, astronomical unit – the average distance between Earth and Sun
BCE, before the Common Era (BC)
BP, before Present
°C, degrees Celsius
CE, Common Era (AD)
CELSS, controlled environment life support system
CHZ, circumstellar habitable zone
CR, cosmic radiation
EM, electromagnetic radiation
ESA, European Space Agency
ESPs, extra-solar planets
ET, extraterrestrial (meaning a sentient organism)
ETI, extraterrestrial intelligence
eV, electron volt
g, acceleration due to gravity
GHZ, galactic habitable zone
Hz, hertz
IR, infrared
K, degrees Kelvin
km, kilometre
l, litre
LED, light emitting diode
LUCA, last universal common ancestor
LW, long wave [radiation]
ly, light year
m, metre
NASA, [US] National Aeronautics and Space Administration
NIR, near infrared [radiation] (the 1,000–3,000nm region of SW)
Pa, pascal (unit of pressure)
PAR, photosynthetically active radiation (400–700nm)
ppm, parts per million
SI, Système Internationale
SETI, search for extraterrestrial intelligence
SW, short wave [radiation] (<3,000 nm)
UV, ultraviolet [radiation] (<380 nm)
v/v, by volume

y, year. This is common usage. The SI unit is "a".

Two useful conversions: 20% O_2 v/v ~= 20kPa; 350 ppm CO_2 (0.035%v/v) ~=35 Pa ; both at sea level.

SI prefixes

Prefixes to form the names and symbols of the decimal multiples and submultiples of SI units:

Multiple	Prefix	Symbol
10^{-9}	nano	n
10^{-6}	micro	μ
10^{-3}	milli	m
10^{-2}	centi	c
10^{-1}	deci	d
10	deca	da
10^2	hecto	h
10^3	kilo	k
10^6	mega	M
10^9	giga	G

1

What is life? Why water?

What is life?

The answer to this first question could almost make a chapter in itself. If we really knew the answer perhaps only a sentence or two would suffice, and the definition would be relegated to the preamble to Chapter 6, which discusses the origin of life. As it is, there is no simple definition; however, we require at least a working hypothesis before we can discuss any of the topics of astrobiology.

The first definition (or more correctly description) of life is informed, coloured, and limited by 'life as we know it' on Earth. It may come as a surprise to many, but there is only one life system here on Earth. This is further discussed in Chapter 6. It is by no means necessary that this is the system of life to be found in different (or even similar) extraterrestrial environments. It is possible to imagine other life biochemistries and forms that evolved under conditions very different from those existing on Earth; indeed, science fiction is full of such creatures! As long as they do not depart from the known laws of physics and chemistry (also an uncertain restriction) such life systems are not completely outrageous. They are certainly often entertaining.

In the 19th century, physical chemists elaborated a short list of laws governing chemical reactions and mechanical work (thermodynamics) on the basis of studies of machines such as steam engines. Ever since, these laws have been found to hold true in engineering, chemistry, and biology. The Nobel prize-winning physicist Erwin Schrödinger, in his seminal book *What is life?*, described the relation between the laws of thermodynamics and life. Although published in 1944, most of what he wrote still holds true today (Dyson 1999). This includes his analysis of the need for information-bearing molecules in living organisms. Schrödinger's perspicacity presaged the revolution in molecular biology of the second half of the 20th century.

The following owes much to Schrödinger. The second law of thermodynamics demands that in any system where work is being done, free energy decreases and entropy (which may be described as disorder) increases. For example, if we place a quantity of sugar in a beaker of water there is considerable order, with the sugar on the bottom and pure water everywhere else. Initially, entropy is said to be low and free energy (the possibility of doing work) is high, but soon work is carried out spontaneously, as the sugar moves by solution and diffusion. Eventually there is a homogeneous mixture of water and sugar molecules. There is no more free energy and entropy is now high (Fig. 1.1).

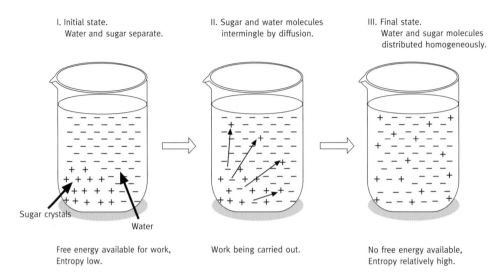

I. Initial state.
Water and sugar separate.

II. Sugar and water molecules intermingle by diffusion.

III. Final state.
Water and sugar molecules distributed homogeneously.

Sugar crystals

Water

Free energy available for work, Entropy low.

Work being carried out.

No free energy available, Entropy relatively high.

Fig. 1.1 Free energy, work and entropy in a beaker of water, with sugar added.

Some 19th-century biologists, looking for a definition of life, thought that life was unique in being able to reduce entropy. For example, when a single fungal spore is introduced into a Petri dish having a layer of apparently homogeneous agar, it soon grows into a maze of fungal hyphae. Similarly, the living contents of a sealed glass jar (which appear to demonstrate a closed ecological system) containing water, algae, higher plants, and fish, may appear to stay alive and grow indefinitely (Fig. 1.2).

The error of the 19th-century biologists was that they did not appreciate that, in the course of time, the second law of thermodynamics holds true in completely closed systems. In the Petri dish system, local energy sources (in this case the agar molecules) will eventually be used up and entropy will increase, as the fungus dies. If the system is open and receives energy from outside, entropy may indeed decrease. The 'ecosystem' in a jar (Fig. 1.2) is, energy-wise, an open system, allowing energy in the form of sunlight to enter from outside the jar. Within the jar, the radiant energy is transduced to chemical energy by the photosynthesis of the algae and higher plants.

These examples are included not so much to describe past errors but to show what is typically being carried out by living organisms. Based on these considerations, some biologists have suggested that life could be defined as a system that transduces energy from one form to another, such as by photosynthesis or consumption of plants by herbivores or herbivores by carnivores. The problem with this definition is that there are many inert, physicochemical, systems which carry out such energy conversions.

Another definition of life, which enjoyed some popularity, is that it is a system following neo-Darwinian rules of evolution. This would include reproduction, storage and transfer of information, and mutation-driven evolution. However, as quickly pointed out, by this definition mules are not alive, while some computer 'viruses' are!

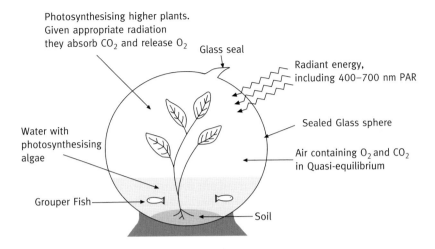

Fig. 1.2 A 'closed', self maintaining ecosystem. PAR = photosynthetically active radiation.

Cleland and Chyba (2002) proposed that the problem of the definition of life is currently unsolvable. They suggest that we need a basic understanding of life, and give the example of water. Until the molecular composition of water was determined as H_2O, there were always confusing examples of liquids, such as ammonia, with at least some similar properties. Cleland and Chyba are no doubt correct in seeking some basic understanding before attempting to define life. However, water may not be the best analogy. It is in fact a very strange molecule. Knowledge of its atomic, molecular, and three-dimensional structure only partially explains its extraordinary properties, which are so essential for life (see below).

Just as for the understanding of the properties of water prior to its molecular and structural analysis, all we can say is that life on Earth is recognizable, but hardly definable, by having all or a large number of the following attributes:

- Life exists in discrete units, usually, but not always, in the form of well-defined cells. Apart from a few, dependent life forms, simple life is monocellular. Advanced life comes in associations of up to many millions of cells acting in concert. Cells are bounded by membranes, themselves essential parts of the living system, that carry out numerous functions. Some of these functions are passive, such as the rejection or attraction of water or other molecules in the exterior milieu. Other membrane functions are active, and involve the expenditure of energy in either denying passage or actively pumping ions and small molecules into and out of the cell against concentration gradients. Almost a century ago A. Oparin in Russia and J. B. S. Haldane in England, two of the pioneers in the study of the origin of life (working independently), defined the primary function of the cell membrane. They considered it essential for maintaining an internal environment different from the external one. The internal environment is more or less constant (in a state of homeostasis) but is never in equilibrium. Although much has been learnt of cells and membranes since the Oparin–Haldane period, their insight has remained central to our understanding and recognition of the very essence of life.

- There is always an energy flow through living systems. This results in local reductions in entropy, and in growth and work. The work may be active pumping into or out of the cell as described above or any other expenditure of energy, such as in mechanical movement or in metabolism, as occurs in our brains as we think.
- Life forms grow and develop in patterns, but unlike most crystals, for example, the final forms are never exactly the same. As discussed below, this very variability in the final forms of apparently genetically identical entities may be a useful criterion for recognizing life.
- Information on construction and operation is formulated and stored. Structures formed are complex and asymmetrical, and both structures and operation are often modified by, and in response to, the environment, which may include the nearby presence of similar or different life forms.
- Units of life, whether of monocellular or advanced form, tend to degenerate over time. This is mainly a result of their continuous exposure to various sources of radiation, toxins, and other environmental stresses. Death is an inescapable characteristic of life. In certain cells of multicellular organisms cell death may also result from genetically programmed, built in, self-destruction mechanisms ('apoptosis'). Consequently, an essential characteristic of life is reproduction and the transfer of information on structure and function to the next generation.
- The transfer of information to offspring is imperfect. There are chance errors. This again may be the result of damage from high-energy radiation or toxic chemical agents in the environment. Such errors are for the most part (more than 99%) detrimental or lethal, and they disappear with the death of the life form. A few survive and result in what are known as trait mutations. Just a very few of these mutations may impart a survival advantage. These positive mutations are rare, but together with activation of dormant genes and rearrangement of genetic material (see Chapter 8), are the very basis of evolution.
- Life forms are composed mainly of liquid water. Water is both a major structural element and the milieu for, and often an integral part of, metabolism. Life is completely dependent on water's unique physical and chemical properties, which differ markedly from those of all similar molecules. Liquid water is absolutely essential to life on Earth. Consequently, conditions where water can and does exist in liquid form serve as the first criteria in the search for possible eco-niches suitable for habitation by life in our Solar System and elsewhere in the universe. Water is so important to our understanding of the functioning of life that it is discussed in detail below.

- Life forms on Earth are built from only 24 of the over 100 known elements on our planet. The most important of these are hydrogen (about 60% by number of atoms), carbon, nitrogen, and oxygen. Each of these 24 elements has properties which appear to be essential for metabolism. They originated from the early nebula of our Solar System, where they constituted just 0.13% of the number of atoms. As described in Chapter 2, not all star systems have even such small complements of these relatively heavy elements essential for life (for a more detailed description of the elements and Earth's structure see Broecker, 1988).

- Apart from water, life's main molecular components are built around skeletons of carbon. Carbon is particularly suitable as a backbone for innumerable and versatile molecules. Carbon atoms tend to catenate (form long chains). The carbon atom easily forms four single-covalent bonds and multiple bonds with other carbon atoms or atoms of other elements. There are more than a million known carbon compounds.

- Life's main operational molecules are proteins, which are constructed from the same 20 of the 100 known amino acids. Amino acids are built of carboxyl and amino groups, with the general formula $COOH–RCHNH_2$. 'R' represents a group from just a single hydrogen atom, as in the simple amino acid glycine, to a heterocyclic ring structure, as in the amino acid proline. Amino acids (or more correctly amino acid residues) can form chains (polypeptides). Strings of polypeptides constitute proteins. They may be most simply classified as being of globular or fibrous form. The enzymes, the protein catalysts of organic life, are of globular form. The fibrous proteins occur in structural materials and may be mechanically active, as in contractile muscle tissue.

- Many asymmetric molecules may appear in one of two mirror-image (chiral) forms. They are characterized by the left- or right-hand rotation of polarized light (L- or D- form, respectively). Amino acids occur in L- or D- chiral forms, but nearly all life forms use the L-form. A rare exception is the presence of amino acids found in some bacteria walls which may be of the D-form. Sugars also occur in L- and D- forms, but living cells use right-hand D-forms almost exclusively. Why life prefers one over the other is not yet understood. There is some evidence that organic material found in carbonaceous meteorites has a bias towards the chirality found in life forms on Earth. This lends some support to the theory that precursor molecules for early life came to Earth from the cosmos during, or shortly after Earth's first billion year Hadean period (Chapter 5).

- All independent life forms contain DNA (deoxyribonucleic acid) the basic building block of the information-storing genes, and RNA (ribonucleic acid) a varied group of molecules that transfer information from the genes, enabling the production of the cell's proteins. A few very primitive life forms, such as most viruses, contain only RNA. However, some part of their life cycle must be passed in more advanced, DNA-containing cells.

- Many life forms have nervous systems of varying degrees of complexity, with which they sense the environment and communicate with each other. The range of development is enormous. It extends from, for example, the single light-sensitive organ of some algae, through environment-sensing higher-plant organs, to neural systems in insects and animals, to the human brain.

- Life forms sometimes move in patterns that serve their function. Again the range of possible movements is very great, from swimming algae through nectar-gathering bees, to fast-running mammals.
- Life on Earth often gives the *appearance* of having been designed for a purpose; a concept called teleology, discussed by Socrates but not accepted today by most scientists (cf. Dawkins, 1986). Life is considered to have evolved by Darwinian selection of chance mutations or rearrangements of genetic material that make the organism better adapted to the ever changing environment.

Ruiz-Mirazo *et al.* (2004) suggest that in addition to the attributes summarized above, life is characterized by its existence in groups of different life forms. They believe that a living organism cannot exist as an individual but is always part of 'the evolutionary process of a whole ecosystem'.

David Wolpert, in a lecture at UCLA in 2005, extended his and William Macready's work on complex systems (Wolpert and Macready 2004) to the recognition of life. He suggested that an identifying and unique characteristic of life is self-*dis*similarity at any given scale. To take his argument to plants: two trees, both grown from vegetative cuttings taken from the same parent tree (and thus genetically identical), may have quite different branching patterns when fully grown. Crystals, in contrast, will grow but usually show great similarity at all scales.

In relation to the above attributes of life and the teleological hypothesis, it is worth mentioning the anthropic principle, as propounded in great detail by Barrow and Tippler (1986; reissued 1996). They showed how life on Earth, including human life, depends on a vast number of fixed physical and chemical constants and parameters— almost as if the universe were designed for life. The dependence of life on these parameters is true enough. However, this does not mean, as they appear to posit, that the universe was designed for life and humans by the mind of a creator. The agreement of the universal constants and our particular environment to the requirements for life could not be otherwise. If there was no agreement we would not be here to contemplate our existence! The anthropic principle argument, when put forward by promoters of the concept of intelligent design, is a logical tortuosity.

The problem with the above list of the attributes of life is that life-mimicking exceptions can be found for nearly each item. We hope to recognize life when we see it, but this can be very unreliable. Consider two examples: science fiction buffs were elated when a rock turned up in an image from a NASA study of Mars (taken by the Viking probe some decades ago) with features similar to those of a human face (Fig. 1.3). Of course this was sheer coincidence, further confounded by our proclivity to see familiar shapes where none exist.

Much more serious, and consequently even more frustrating, was a mineral form inside a meteorite found on Earth (Fig. 1.4). The composition of the gases enshrined in

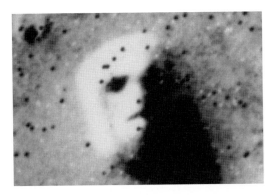

Fig. 1.3 A sculpture of a human face on Mars? Courtesy NASA/JPL/Malin Space Science Systems.

this meteorite is accepted as evidence that it originated some billions of years ago from Mars. It appeared to contain fossilized, worm-like, segmented bacteria that existed on Mars at a time when it may have had a thick atmosphere and liquid water on its surface. The original announcement of this finding by NASA scientists (in 1996) was circumscribed by clear warnings of the tentative nature of the conclusions. This did not prevent the news media from publishing exaggerated reports on the finding of extra-terrestrial life. Later analysis suggested that it was unlikely that the 'fossil forms' found in the meteorite resulted from the activity of living creatures.

The main argument against the apparent fossils in the Mars meteorite representing once living forms is based on size. A minimum quantity of the common molecules of life (DNA, RNA, and proteins) requires a volume contained in a sphere of about 200 nm diameter. This is many times the size of the apparent bacterial forms in the Martian meteorite. Most Earth bacteria are many times larger. Some much smaller 'nano-bacteria', which have been identified on Earth, are probably dependent entities. Essential parts of their life cycles must be spent in other, larger, life forms, such as other bacteria or animal guts. Moreover, some inorganic minerals may take such truncated forms. This Martian meteorite finding is more than a decade old, but the dependence of all nano-bacteria on larger life forms is still uncertain and truncated mineral formations are rare, so the debate continues (e.g. Knoll 2003, Chapter 13).

The Earth is often thought to be teeming with myriads of life forms. And indeed, if we emphasize the word 'form', it is. However, they are all made and seem to have evolved from the same basic building blocks. In other words there seems to be only one *type* of life on Earth. To date even the most primitive organisms have been found to contain at least RNA and usually also DNA. Why no other basically different life forms have appeared, or survived, on Earth is discussed in Chapter 6 (but the question is not satisfactorily answered). For example, there seems to be no organism built from amino acids with D- chirality, or having five instead of the ubiquitous four bases (always the same ones) in its genetic code. Even organisms from the most exotic habitats on Earth,

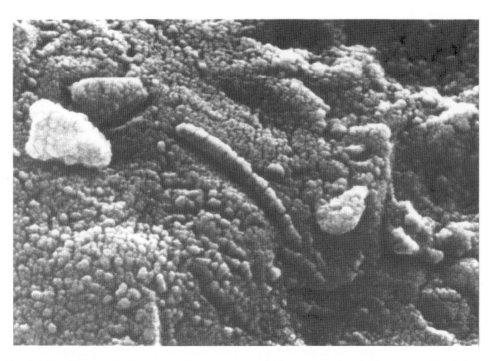

Fig. 1.4 Fossilized, microbe-worm-like structure, within a meteorite which came to Earth from Mars and was discovered in Antartica in 1984. Courtesy of NASA.

such as bacteria from 3 km below the surface of the Earth, from around hot sulphurous springs, or higher life forms in the strange ecosystems around deep sea vents (Chapter 9), have the same basic structure, with information storage and handling based on DNA and RNA.

Consequently, a very definite criterion for identifying life on Earth is the presence of RNA. Unless we have been missing the 'aliens among us' (see Chapter 6 for a discussion of this possibility) this yardstick would almost certainly work on Earth. It would be surprising if this also holds true for life forms that have evolved elsewhere. A comprehensive summary of criteria for identifying extraterrestrial life, but always 'life as we know it', is given by Lunine (2005, Chapter 14).

In conclusion, a definition of life still eludes us. This is not a trivial problem when we study the conditions required for the appearance of life on Earth, or search for extraterrestrial life.

Why water?

Viewed from afar, planet Earth seems to be a planet consisting mainly of water. Water covers 71% of its surface. Water on Earth is mainly liquid, but some is in the solid ice phase. Despite appearances, most of the Earth is actually composed of either solid elements or liquid metal. Water is mainly near or on its surface. Geologists still argue as to whether this water came from deep within the Earth (where some may continue to exist) or from extraterrestrial sources during the first billion (10^9) years of Earth's formation (Chapter 4). A small percentage of Earth's water is in the atmosphere as gas water vapour, or as liquid cloud droplets. The total liquid water on Earth has a volume of about 1380×10^6 km^3. Most of this (97%) is in the oceans which contain some 3% dissolved salt. Apart from the bacteria present to a depth of 3 km within the Earth, most life in the biosphere is in the oceans. A further 24×10^6 km^3 (2.4%) of the Earth's water is in the solid phase, frozen in the snow and ice of the Arctic and Antarctic regions and in glaciers. If converted to liquid water, the water in the atmosphere would fill a volume of some 13×10^3 km^3.

Given the prominent presence of water on the surface of Earth it is really not surprising that early life, and all subsequent life forms that developed on Earth, were and are based on water. All life exists in an environment of water. It constitutes some 50% of all life forms, although this may vary from just a few per cent in dormant seeds to close to 98% in most algae. Water forms part of the cell structure while also serving as the main solvent. It is a vital reactant or product in many metabolic processes, such as photosynthesis and animal and plant respiration.

Terrestrial life (meaning in this case life on dry land, which includes of course humans) is reliant on so-called fresh water. 'Fresh' water is arbitrarily defined as having a salt content of less than one-third that of ocean water. It represents only 2.8% of Earth's total water. Aquifers below but relatively near the surface of Earth contain some 23×10^6 km^3 water, of which only about half may be described as fresh. Surface water forms 0.022% of the total. Of the latter, less than 0.3% is fresh. About the same volume of surface lakes are saline. As discussed in Chapter 10, fresh liquid water is already a precious and dwindling resource in many regions of the world inhabited by humans. Note that the polar icecaps and glaciers are also formed from essentially salt-free water. They contain 77% of Earth's fresh water.

Despite being such a common molecule on Earth, water is extremely unusual in its chemical and physical properties. This is particularly notable when it is compared with other molecules of similar size and mass (Fig. 1.5). Life, which evolved on our watery planet adapted to and became entirely dependent on these unique characteristics. Some of the special properties of water are described below.

Water as a solvent and structural component

The ionic charges of water are separated in space (Fig. 1.6) which makes it a very polar liquid. When other ionic or polar substances are immersed in water, the water molecules tend to surround (or in other words dissolve) them. The polarity of water is sometimes exploited by living cells to promote partitioning, for example of the hydrophobic, lipid-walled mitochondria (Pohorille and Wilson 1995). Mitochondria are organelles within cells which carry out most of the process of respiration, which is inhibited by water. Water does show a small degree of solubility for non-polar substances such as oils and fats, making it an excellent general solvent for biological processes.

Water does not actually dissolve many large molecules, such as proteins and DNA, but it adheres to and surrounds them. This may be of great importance. Many of the operational characteristics of these large molecules depend on the spatial relationships of their active sites. The adhering water contributes to the mechanical stabilization of these large, often filamentous, macromolecules. This adhering water may be more solid, in an 'ice' matrix (see below), than liquid. It attaches by hydrogen bonding to amino acids and proteins for example. In addition, strings of water molecules (see below) have

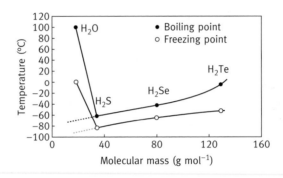

Fig. 1.5 Boiling and freezing points of water and some molecules of similar molecular weight at a standard pressure of 1 bar (the atmospheric pressure at sea level).

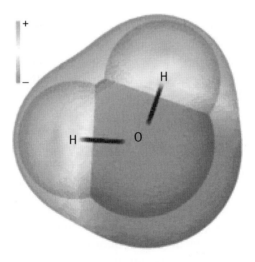

Fig. 1.6 Structure of the water molecule.

the special property of rapidly transmitting H^+ and OH^- ions, which is an essential part of metabolic activity.

Another important property of water is that it is amphoteric. This means that it is able to act as either an acid or a base. Most cellular metabolism occurs at a pH close to neutral (7.0), although small deviations from this value may be of considerable importance.

Boiling and freezing points of water

As shown in Fig. 1.5 the boiling and freezing points of water are very different from those of other similar sized molecules. The data presentation in this figure has been ascribed to Linus Pauling, one of the great chemists of the 20th century, who clarified many of the special properties of water. It clearly shows how in comparison to other hydride molecules of similar size and mass, water has anomalous boiling and freezing points (100°C and 0°C, respectively). These two values are themselves of utmost importance to life, as the surface equilibrium temperature of Earth (today averaging ~15°C) is in this range in which water is a liquid. Moreover, throughout the last 3.7 billion years of life's history, the temperature has remained within this range at least somewhere on Earth. The perpetual presence of liquid water has enabled the continuity and evolution of life. This was the situation despite the large physical changes that have occurred in the environment. The latter included a ~20% increase in solar radiation, extreme variations in the composition of the atmosphere, changes in the surface distribution of oceans and land masses, and recurrent cataclysms resulting from ice ages, volcanism, and the impact of comets, giant meteorites and bolides. These changes, and how this rather amazing temperature stability was maintained, are discussed in later chapters.

Specific and latent heats of water

Water has a high specific heat (defined as the energy required to raise the temperature of 1 g of a substance by 1°C) and high latent heats of freezing and boiling (defined as the energy required to produce a phase change, between liquid and solid or liquid and gas, without a change in temperature). Table 1.1 compares these properties with those of some common and similar liquids.

As seen in Table 1.1, water has a relatively high specific heat (only a little less than that of ammonia) and very high latent heats of both freezing and boiling. Consequently, water serves as a temperature buffer. Life requires this stability, as enzymes (the essential catalysts required for metabolism) have defined temperatures for optimum function. For example, leaves may dissipate much heat by the evaporation of only a little water in transpiration; or animal body temperatures only change when a large amount of heat is either absorbed or lost by radiative or sensible heat exchange or by evaporation (assuming no heat production by metabolism).

The same high heat buffering capacity of water tends to stabilize the temperatures of the oceans, and in turn the climate on the Earth's surface. Such stability is essential for the more developed life forms. Great ocean currents move from equatorial regions to higher latitudes. The heat capacity of these waters moderates the climate of some coastal regions, making them suitable for habitation by many forms of life, including humans.

The dielectric constant of water

Many basic chemical and biochemical reactions take place between ions and charged active sites of larger molecules. For example, common salt, the quite inactive NaCl, dissociates in water to give the very active Na^+ and Cl^- ions. The degree of dissociation is determined by the solute and by the dielectric constant of the solvent. The higher the dielectric constant the greater the ionic dissociation and the possibility of metabolic reactions. The least dissociation would be in a vacuum where the dielectric constant is given the arbitrary value of 1. As shown in Table 1.2, apart from a few solvents and solutions which are otherwise unsuitable for life, water has a relatively high dielectric constant. Again, this makes water a very suitable milieu for biochemical processes.

Molecular and spatial structure of water, and hydrogen bonds

Many of the anomalous properties of water, on which life so depends, may be at least partly explained by what we know of its molecular and spatial structure. The structure of a single water molecule is shown in Fig. 1.6. As shown in this figure, the water molecule has two positively charged hydrogen atoms on one side and a negatively charged oxygen atom on the other. This and the geometry of the molecule are the reasons for its polarity.

Water molecules are attracted to each other and to other similarly charged molecules by hydrogen bonding. Hydrogen bonds are formed between hydrogen and other atoms of opposite polarity, typically oxygen. The exact nature of the bond is not known, but it is considered to be mainly a result of electrostatic forces. One hydrogen may be attached to up to four atoms of other molecules. The strength of hydrogen bonds (~18.8 kJ mol^{-1}) falls between the low strength of van der Waals and hydrophobic bonds (about 2–6 kJ mol^{-1}) and the higher strength of ionic and covalent bonds (varying between 4

Table 1.1 Specific and latent heats of water and some common liquids

Substance	Specific heat ($J\,g^{-1}$)	Latent heat of fusion (melting) ($J\,g^{-1}$)	Latent heat of vaporization (boiling) ($J\,g^{-1}$)
Water	4.19	335	2,260
Ammonia	5.15	108	1,368
Ethanol	2.51	105	854
Benzene	1.75	1.26	393

and 400 kJ mol^{-1}). This intermediate value is of some importance. It is high enough to maintain a bond, but not so strong as to prevent it being broken, as in the change between ice and water described below.

In photosynthesis, the intra-molecular covalent bonds of water molecules are split by energy obtained from the Sun's radiation. The latter is at wavelengths between 300 and 3000 nm. At wavelengths below 720 nm the discrete units of radiation, the photons, have sufficient electrical potential energy, >1.8 eV, to excite the photosynthetic pigments and drive the photosynthetic mechanism (see Chapter 6). However, only about 48% of the Sun's radiation is in this waveband. As a result, photosynthesis can only utilize about half of the available solar energy. The emphasis here is on 'can'. Plants actually use much less of the available radiation (at most 4–8%). This is discussed in detail in Chapter 6.

The chlorophylls, together with a number of auxiliary pigments of oxygenic plants, absorb most of the radiation between 400 and 720 nm, but not higher wavelengths. This is often considered to be the limiting factor in the use of solar radiation by oxygenic photosynthesis. In a way this is true. Certainly there can be no photoreaction without a photon receptor.

As discussed in Chapter 6, the first evolutionary 'attempts' at photosynthesis, utilized radiation at wavelengths longer than 720 nm. This was possible as sulphur-containing (and some other) molecules were exploited as electron acceptors. Their basic molecular bonding could be split by relatively low levels of energy. Such molecules soon became rare on the Earth's surface. In the course of evolution, photosynthesis adapted to exploit the freely available water molecule, which could be used as an electron acceptor. However, the splitting of water requires photons of a higher energy level (>1.8 eV) only found in photons with wavelengths shorter than 720 nm. These photons are absorbed by the chlorophylls and auxiliary pigments. So, it is not pigments which limit the use of sunlight but rather the lack of a weaker-bonded electron acceptor. Resources, especially the often deficient nitrogen, are conserved by not having pigments which absorb at wavelengths which cannot be used. Leaves of plants have evolved to reflect radiation >720 nm. This reduces futile water loss, resulting from overheating.

Table 1.2 Dielectric constants of water and some common liquids and solutions

Substance	Dielectric constant
Octane	1.96
Benzene	2.28
Acetone	21.4
Water	80.4 (at 20°C)
Hydrogen cyanide	116.0
Glycine	137.0

Data on Hydrogen cyanide and Glycine are for solutions in water.

As a result of the hydrogen bonding between water molecules, water can form hexagonal lattice structures. This is the form in which water exists as ice (see Fig. 1.7).

The degree of structure in water varies. In ice it is 100% crystalline but some degree of structure remains in liquid water. This 'clustering' of water in groups of 14 to as many as 280 (and possibly more) molecules explains many of its anomalous characteristics and its interactions with protein molecules. One important characteristic of water which can be explained by clustering is its maximum density at a temperature a little below 4°C. The open lattice ice structure has a lower density than liquid water. As the temperature of water rises above the freezing point (0°C for fresh water and −1.9°C for sea water) two opposite tendencies come into play. First the structure collapses, increasing density, while at the same time the increased kinetic energy, resulting from the higher temperatures, pushes the molecules apart and thus decreases density. From 0 to 3.9°C the structure collapsing effect dominates, while above this temperature the kinetic effect reduces density. Consequently the greatest density of water is not, as for most liquids, at it freezing point, but a few degrees higher.

This phenomenon has strongly affected life in past aeons, and continues to do so today. As described in later chapters, there have been many periods (ice ages) when large parts of the Earth were covered in ice. Even today the region of the North Pole is mainly an ocean covered in ice. Moreover, in the winter season many northerly and southerly lakes and rivers are frozen. In nearly all these times and places, liquid water, at temperatures between 0 and 3.9°C, remains below floating ice. The ice even provides some degree of thermal insulation. This phenomenon enables life to survive below ice even in extremely cold conditions, such as in the Arctic Ocean. As described in Chapter 4, this raises exciting possibilities for the existence of life on the planet satellites Europa and Enceladus, both of which probably have oceans below a water-ice covered surface.

Life has either adapted to, or exploited, many of the special characteristics of water caused by its partial multimolecular structure, even at temperatures well above that of its apparent freezing point. A classic example is the rise of sap in tall trees. It was long a puzzle as to how water could rise to heights greater than 10 m. Up to this height, sea level atmospheric pressure on the root water could raise the water in the plant pipes (xylem vessels) if it was removed from the top of the plant by evaporation. For plants taller than 10 m, water should rise in the xylem vessels to 10 m but above this height there would be a vacuum, or the vessel would fill with penetrating air. There would be no problem if the water was pumped from below, by root action, but there is no such active pumping. The solution to this problem is that the cohesion between water molecules is strong enough for them to behave like links in an interlocking chain. The chain is raised by being pulled from the top. To express this 'pull' in more physically rigorous terms, the evaporation of water in the leaves reduces the water potential in the leaf to below that in the soil. This produces a hydraulic potential gradient from the soil to the top of the plant. There are other complicating factors, but the system works in the narrow vessels of tall trees and vines, which sometimes reach a height of >30 m.

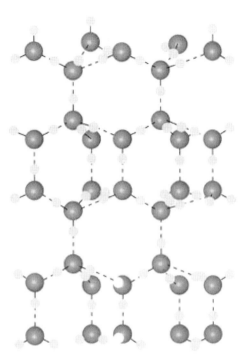

Fig. 1.7 The lattice, ice structure of water.

Another phenomenon easily explained by the interlocking structure of water molecules is surface tension. This results from the monomolecular film of tightly bound molecules which forms on the surface of water. Some very small insects are able to walk on this film in search of prey. The same phenomenon also explains water's high latent heat of evaporation.

Discussion

This section began with the question 'Why water?'. It rather begs comparison with the 'Which came first, chicken or egg?' conundrum, except that in this case we know that water came before life. This means that life had to adapt to and then, in the course of evolution, learn to make use of the special properties of water. So, when we assume that water is absolutely essential for life, we must again and again remember that we are limited by the caveat of life as we know it, on planet Earth.

When astrobiologists search for extraterrestrial life they put the presence of conditions of temperature and ambient gas pressure that support the presence of water (and liquid water in particular) as primary search criteria. However, they recognize that this is looking for Earth-similar life. There is no a priori reason to believe that life could not evolve in other liquids, such as methane. Methane is a liquid at temperatures between −164 and −183°C. Traces of methane gas have been found in the atmosphere of some extraterrestrial planets and low temperatures are common on planets more distant than Earth is from the Sun. Moreover, studies of Saturn's satellite Titan, supported by the recent (2004) findings of the Cassini–Huygens Titan probe, indicate that it has lakes or small oceans of methane. Recent discoveries of bacterial life and complete ecosystems in the dark and cold around deep sea methane vents on Earth show that photosynthetic radiation (in our case from the Sun) is not the only possible energy source for life. Certainly such life may be very very different from what we know on Earth. Allowing ourselves some science-fictional speculation, perhaps the biochemical reactions of life in methane at −170°C would be so slow and lifespan so extended as to make star travel possible for the inhabitants of Titan? However, and 'to avoid all doubt' as the lawyers say, to date (2008) no life has been detected on Titan, or anywhere else, other than Earth.

Summary

Defining life

Life, even with the qualification 'as we know it on Earth', is extremely difficult to define. There is no one satisfactory definition. Generally life is a molecular system which conforms to or does most or all of the following:

- In addition to the ubiquitous hydrogen, it contains certain heavy elements such as oxygen, iron, and carbon, which are not present in every star system.
- It is based on water in the liquid phase and dependent on the special properties of water.
- It is formed from basic molecules of specific configurations (such as amino acids having left-handed chirality).
- It forms cellular units bounded by membranes and maintains an internal milieu different from that outside.
- It transduces and expends energy, producing work, growth, and local reductions in entropy.
- It conserves information on structure and function.
- It reproduces and dies after transferring structural and functional information to the next generation.
- Information transferred has occasional, random, usually detrimental errors. These errors, when they survive, are called mutations. When advantageous they, and rearrangements of existing genetic information, are essential for evolution.
- It always contains RNA and nearly always DNA.
- It usually appears in groups or ecosystems.
- It is characterized by the *dis*similarity of individuals at different scales of development.

Water is a very special molecule

Although water is common on and below the surface of Earth its physicochemical characteristics are unique in comparison with those of similar molecules. In the course of evolution, life has adapted to and exploited these properties, which include:

- It is a universal solvent, especially for polar substances.
- It is amphoteric, i.e. can act as either an acid or a base.
- In the liquid phase at temperatures prevalent in most places on the Earth's surface:
  - it forms an open lattice structure in ice, which is not entirely broken at temperatures above freezing point;
  - it has its greatest density at a temperature above freezing, enabling life in water below a cover of ice;

- it has a high temperature-buffering capacity—important from the climate scale to the scale of maintaining the temperatures of animals and plants;
- it has a high dielectric constant, facilitating ionic reactions;
- it solvates ions and also macromolecules, contributing to the latter's structural stability and to the rapid transport of protons.
- Life on Earth is so intimately connected to the special properties of water that the presence of water and physical conditions which could allow water to be in its liquid phase, serve as prime criteria in the search for extraterrestrial life.

2
The universe from the perspective of biology

Life exists in the universe; humans and other life forms on Earth are all the evidence required. The central theme of astrobiology is to learn whether we are alone, or are only one of many other life forms out there, some perhaps intelligent. In the last three or four decades opinion has been divided between optimists ('we are not alone') and pessimists, who believe that the chances of finding extraterrestrial life are extremely small. These are emotion-laden terms. The optimists yearn for the possibility of exchanging knowledge with other sentient beings. Others think we should be lucky to find that we are indeed alone, and not molested by superior, exotic aliens, who may not be benevolent, or vegetarian!

The 'optimists' have usually come from among astronomers, who have an appreciation of the vastness of the universe. For them, the sheer number of stars and galaxies makes it unlikely that Earth is the only place where life has evolved. The 'pessimists' have usually been biologists. They are more aware of the enormous difficulties and the countless requirements that must be satisfied before life can appear, persist, and evolve to the level of *Homo sapiens sapiens*. This latter concern is a central motif of the astrobiology of Earth. Biologists point out that only once in the 4.55 Gy of Earth's existence has a life form appeared on Earth capable of seeking out other extraterrestrial life: 4.55 Gy is more than a third of the lifetime of the universe. Moreover, the appearance of *Homo sapiens sapiens* was a *very* recent chance event, which might not have occurred at all! These opposite views were elegantly argued more than a decade ago by the astronomer Carl Sagan and the biologist Ernst Mayr, in open debate in newsletters of the Planetary Society (Sagan and Mayr 1955, see list of internet portals in References and Resources; see also Impey 1995).

The search for life, and intelligent life, in the universe is discussed in more detail in Chapter 10. This is in relation to the possibility of humans escaping from an Earth of dwindling resources by exploring and colonizing our own near cluster of stars, the Milky Way galaxy. At this point it is interesting to note that the role of the 'optimists' and 'pessimists' has recently begun to reverse. On the one hand biologists have been deeply impressed by the discovery of numerous extra-Solar System planets (see Kerr 2008). In addition, within our Solar System, a number of niche environments have been found where even some Earth extremophiles could possibly exist (see Chapter 8). On the other hand, astronomers are becoming more aware of the fastidious and uncompromising requirements for life, which are not satisfied in many types of galaxies, in vast regions of our own galaxy, in most star systems, and on most planets. (The term 'galaxy' is further explained below.)

Today many, perhaps most, astrobiologists believe that there are probably numerous occurrences of simple, unicellular life on other planets in the universe and possibly even in the Solar System. However, complex, multicellular, and certainly intelligent life is much more unlikely. There is certainly little chance of finding extraterrestrial intelligent life in our Solar System, a view quite different from that prevalent in the late 19th and early 20th centuries. At that time some astronomers thought they could see evidence of a past civilization on Mars. Moreover, given our present level of science and technology, two-way communication with intelligent life on a planet orbiting another star is hardly feasible. Human space travel to a planet orbiting even our nearest star neighbour will certainly not be possible in the foreseeable future.

The terms universe and cosmos are really one and the same, meaning the entire known 'world': planets, comets, stars, galaxies, life, humans, everything. 'Cosmology' is the study of the entire universe. In the USA and western Europe, pilots of space vehicles are called 'astronauts'. The Russians prefer 'cosmonauts'. The truth is, both names are hyperbole. No human has ever ventured further than the Moon. In terms of the universe, that is just the Earth's backyard.

This chapter sets out what is known of the contents and structure of the universe from the special perspective of biology. The stage is set for understanding where in the cosmos conditions for life may exist, and what special circumstances in our local corner of the universe allowed life to appear and persist on Earth.

There may be other forms of life elsewhere in the cosmos, even intelligent life, living under conditions quite different from those in our biosphere. In the absence of knowledge of such life forms, the search for environmental conditions that may support life is by necessity informed and coloured by the life with which we are familiar. As discussed in Chapter 1, a major and essentially unresolved problem is to define exactly what we mean by 'life', even as it appears on Earth. Here the term 'life' is used without an exact definition, but within the general caveat of 'life as we know it on Earth'.

The cosmos is studied today mainly with telescopes that use the many different wavebands of the electromagnetic (EM) spectrum. These include: the 350–750 nm region which contains radiation visible to the human eye; wavebands shorter than 300 nm, which may be so energetic as to destroy life; the near infrared (700–3,000 nm), as given off by bodies at surface temperatures close to that of Earth ($\sim$300 K) and longer wavebands. The latter are emitted by certain relatively cool cosmic bodies and are also used for radio communication. In order to better observe other galaxies and bodies outside our galaxy astronomers rely on wavelengths that are not strongly absorbed when they traverse our galaxy, the Milky Way. The wavebands which tend to be absorbed by the dust of the Milky Way are in the ultraviolet (<350 nm) and visible (400–750 nm) regions of the spectrum. The many different radiations (including non-EM), their sources, and significance to life are discussed in chapters 3 and 5.

Some special units of measurement used in astronomy

The SI (Système Internationale) system of units is the one adopted by modern science. In this system 1 billion is defined as a thousand million (10^9). (Note that the British billion was formerly defined as 10^{12}: this is now obsolete in scientific and most general usage.) The prefix in the SI system denoting 10^9 is 'G' and the official symbol for year (which is not an SI unit) is 'a', thus a billion years is correctly denoted 1 Ga. However, 'a' is rarely used and the abbreviations 'yr' or 'y' are common. Note that the enormous distances and times in the universe are so much greater than those we commonly encounter on Earth that astronomers have formulated more practical units of measurement.

It is simply too cumbersome to use kilometres for astronomical distances. Instead the 'light-year' (ly) is used here. According to Einstein (and all but the most esoterically thinking cosmologists) light travels at the fastest speed possible, which is almost 300,000 km s^{-1}. Using units based on the speed of light, distances are defined and measured as the time light takes to travel a certain distance. For example, Proxima Centauri, the star nearest to our Solar System, is at a distance of 4.3 ly. This means that a beam of light (or any EM radiation, such as a radio signal) would take 4.3 years to travel between Proxima Centauri and our Sun.

The Milky Way, the galaxy (see below) which is home to our Sun, is so large that it would take a beam of light some 100,000 years to travel from one edge to the other. (Estimates, based on different criteria for defining the edge of the galaxy's halo—a cloud of old stars, dust, and dark matter—vary between 80 and 150,000 ly.) Light reaching our Solar System from the furthest galaxies took more than 12 Gy to reach us. These vast distances, together with the *relative* slowness of the speed of light, have significant consequences for humans. For example, let's assume that a 'close by' civilization signals us by radio from a star system at a distance of 1,000 ly (just one-hundredth of the breadth of the galaxy). Even if we replied immediately, it would be 2000 y before they received our response. By that time they, and we, may have ceased to exist.

The 'AU' or 'astronomical unit' is used for the more 'modest' distances within the Solar System. It is defined as the average distance between Earth and the Sun. One AU is roughly 150 M km, or 93 M miles.

Another unit of distance which is more common in the astronomy literature is the 'parsec' (short for 'parallax second'). It is based on the parallax phenomenon of relatively close-by stars as seen from Earth. Note that 'second' refers here to the angular unit and not the time unit. By simple geometry, a parsec may be equated to about 3.62 ly.

Table 2.1 Some cosmological distances (all approximate)

	In kilometres	In light years (ly) or Astronomical Units (AU)
Most distant detectable galaxies	1.4×10^{23}	1.3×10^{10} ly
Distance to the Sagittarius-Dwarf galaxy (nearest to the Milky Way)	1.86×10^{18}	8×10^{4} ly
Diameter of the Milky Way[a]	10^{18}	10^{5} ly
Sun to Alpha Centauri (star nearest to the Sun)	10^{14}	4.3 ly
Distance from Sun to the Oort comet cloud	7.5×10^{12}	5×10^{4} AU
Diameter of the Solar System (to outer planet and Kuiper asteroid belt)	$(4.5–7.5) \times 10^{9}$	30–50 AU
Average[b] Earth–Sun distance	1.496×10^{8}	Defined as 1 AU
Diameter of Earth	12.756×10^{3}	

[a] Depends on the definition of the galaxy's boundary. The Milky Way has a 'halo' which may extend to hundreds of thousands of light years. By this definition the Sagittarius-Dwarf galaxy is within the Milky Way.

[b] See chapters 4 and 5 to learn why an "average" and not a fixed distance.

To get some appreciation of the need to express astronomical distances in light-years, try writing the distance across our galaxy in kilometres: multiply 100,000 light-years × 365 days × 24 hours × 60 minutes × 60 seconds × 300,000 km s^{-1}!

In Table 2.1, using the above units and kilometres for comparison, we put some approximate numbers to distances in the universe.

Time

If the term 'life' defies definition 'time' is even more difficult. Philosophers and scientists have argued its meaning since *time* immemorial—as you see it is difficult to avoid using it. At best, time scales can be constructed between different events which occurred in the past, occur now, or will occur in the future. They are expressed in terms of recurring events, such as the rotation of the Earth on its axis, giving us the day, or the annual orbiting of Earth around the Sun. Some of these time scales are shown in Table 2.2.

Life is driven by biochemical reactions, such as the splitting of plant starch into digestible sugars. Some of these reactions take seconds. Our lives are also affected and effected by events as short or shorter than milliseconds, such as the splitting of a water molecule by light in plant photosynthesis, or events as long as the year-long circumnavigation of Earth around the Sun. Taking the long view, life has been affected by events in the cosmos since its inception on Earth, 3.7 Gy BP.

It is possible to predict the ultimate death of our Sun, when the source of its energy is exhausted. This will be in some 4–6 Gy from now.

Table 2.2 Timing of some events affecting life, on Earth and in the universe

Recurring events	
Earth rotates on its axis	1 day
Earth orbits the Sun	1 y
Comet Halley orbits the Solar System	76 y
Sun revolves around the centre of the Milky Way	~250 My
Maximum lifespan of a human being	~120 y
Historical events (all dates are approximate)	
Appearance of human technology capable of seeking extraterrestrial life	Last 50 y
Last, documented major impact of Earth by an object from space (the Tunguska event in 1908)	100 y BP
First (and to date last) observed impact of a comet on planet Jupiter	~15 y BP (in 1994)
End of last ice age	11,000 y BP
First 'intelligent' humans (*Homo sapiens sapiens*) appeared. (Palaeobiologists argue the exact timing; cynics say that this event has not yet occurred!)	100,000 y BP
Appearance of first humanoids	5 My BP
Last great extinction event (including that of dinosaurs)	65 My BP
First multicellular life forms appeared on Earth	600–500 My BP
First photosynthetic, oxygen-evolving life, on Earth	2.5 Gy BP
First life appeared on Earth	3.7 Gy BP
Earth formed in the Solar System	4.55 Gy BP
Early galaxies formed	12–13 Gy BP
The Big Bang; beginning of our universe and time	13.7 Gy BP
Sun will complete its nuclear cycle	4–6 Gy in the future

There is one unique (and for humans usually unfortunate) characteristic of time. Unlike other vectors on land or in space, time appears to be irreversible. We were all born in the past, exist today, and will die in the very near future. Unless there is some stupendous medical breakthrough, today's 6 G humans will all be dead by the year 2150—just a flicker of time on the astronomical, geological, or evolutionary time scales. As discussed in Chapter 10, our short life span is a major constraint on human navigation of the cosmos.

Biologically significant events occurring in the universe

The most widely accepted reading of Einstein's theory of relativity implies a universe with a definite beginning. Building on his equations and extrapolating backward in time suggests that the universe began 8–20 billion years ago. According to this, and calculations based on observations of the constantly retreating galaxies, the entire mass and energy of the universe were initially concentrated in a single, discrete spot of infinite mass and energy, which physicists call a 'singularity'. At a certain moment this singularity exploded (the so-called 'Big Bang') giving rise to the present universe. At that very moment time began. Recent measurements by a NASA probe of the 'fossil radiation' background in space, which is a remnant of the Big Bang, has supported this theory and allowed dating of that seminal event to close to 13.7 Gy BP.

Compared with the human life span, 13.7 Gy appears to be an extraordinarily long time, but from the point of view of life on Earth it is not. Life has existed on Earth for more than a quarter of this period, about 3.7 Gy. However, humanity only appeared after nearly 3.7 Gy of evolution, and only in the last centuries has modern science and technology developed. Even today humans (or at least their technology) have not yet evolved to a level which would allow them to communicate and travel to other civilizations in the galaxy (assuming there are any).

Most famously (especially for those of us who grew up in fear of nuclear war) Einstein showed how energy and matter were essentially interchangeable. The data from the NASA background radiation probe also resulted in some unexpected values for the energy and mass composition of today's cosmos (Fig. 2.1). The two major components turn out to be dark energy (73%) and dark matter (23%). Hydrogen and helium form just 3.5% of the total. The term 'dark' indicates that they do not emit radiation that can be seen. Their exact nature, or even existence, are still (in 2009) uncertain. Stars constitute only 0.5% of the universe. 'Metal' elements, an astronomical term for those atoms or ions heavier than helium, such as carbon, nitrogen, and iron, constitute only 0.03% of the total mass. This is of considerable significance for biology. In addition to hydrogen, life is constructed of such metalloid elements. Moreover, on Earth, higher multicellular life forms depend directly or indirectly on energy coming from the Sun. This paucity of heavy elements (in astronomers' meaning of the term) and available energy put severe limits on the distribution of life in the universe.

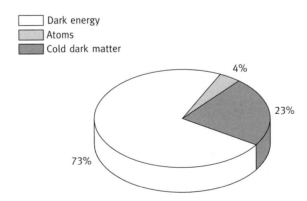

Fig. 2.1 Composition of the universe as derived from measurements of the residual radiation from the Big Bang.

Stars

The first stars appeared some 200 million years after the Big Bang. On the scale of 13.7 billion years of cosmic history this is very soon after the beginning of time. The earliest stars were formed of hydrogen and helium only. All stars derive their enormous energy from nuclear fusion. At first this is fusion of hydrogen nuclei, which produces helium. When the hydrogen is exhausted helium is fused, resulting in the production of the heavier elements. In massive stars the nuclear furnace stops when the line ends with the fusion of silicon and the production of iron. As noted above, and unlike chemists, astronomers call all these heavier elements 'metallic'.

The number of stars is very, very large. To get a rough idea multiply the number of stars in an average large star cluster (galaxy), which is in the hundreds of billions, by the estimated number of galaxies—a similarly enormous number. Stars are classified by a number of different criteria. The first puts them into seven groups, characterized by their surface temperature. The Sun is in the fifth group (group G) with a surface temperature of about 6,000 K. The surface temperature is of great importance to life as it determines the radiation spectrum. Stars are then divided into groups based on luminosity. In this case the word 'luminosity' refers to the total radiation emitted by the star, not just the visible waveband. Our Sun belongs to the fifth and largest 'main sequence' class. These stars are relatively small and young and have a lifetime of at least 5 billion years. Advanced life on Earth is entirely dependent on the intensity of solar radiation, its spectrum (mainly, but not only, for photosynthesis and vision), and longevity (required by evolution). Stars of great mass, for example, could not have life on their planets. They have surface temperatures so high that they radiate strongly in the shortwave bands of the spectrum. Such short-wave radiation (like the small percentage of ultraviolet (UV) radiation in the Sun's spectrum) damages or destroys living cells.

About half of all stars are binary, meaning that there are actually two stars which rotate around a common centre of mass. Any planets they have would either be large and closely orbiting one of the stars or orbiting both, at a vast distance. It is also possible that other planets, initially associated with the binary system, would be completely ejected.

None of these scenarios could produce a planet with an environment compatible with life.

When stars of great mass (between 8 and 25 times that of our Sun) complete their nuclear cycle, they first implode. They then explode and form what are called supernovae. When this occurs, vast amounts of matter and radiant energy are ejected into space. This is both good and bad for life. The good news is that a supernova produces the heavy 'metallic' elements from which planets and life are made. There is evidence that a shock wave from a supernova triggered the formation of our Solar System some 4.55 billion years ago. The bad news is that if life already exists in a nearby star system it could be destroyed by the intense burst of proton particle, X-ray, and gamma-ray radiation coming from the nearby supernova explosion. Such radiation would rapidly destroy the ozone layer in Earth's upper atmosphere, exposing the surface to life-destroying solar UV radiation. The shorter-wave EM radiation would then destroy surface life.

In the lifetime of our own galaxy, the Milky Way, there have been some 100 million supernovae explosions. Mercifully, throughout the entire 3.7 Gy history of life on Earth none has been close to our Solar System. The last supernova in the Milky Way was seen in 1680. In case we need to be reminded of our vulnerability, a supernova explosion was observed quite recently in the Magellan Cloud, a small galaxy near the Milky Way (Fig. 2.2). Supernovae have long been noted by astronomers in Europe, China, and ancient Babylon, as far back as 3,000 y BP. Their recordings noted the appearance and then disappearance of 'very bright stars' (not of course of 'supernovae'). Recently, the first observation has been made of a supernova in its very first, initial X-ray outburst (Soderberg *et al.* 2008).

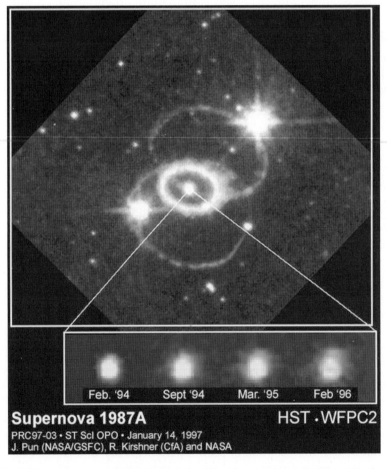

Fig. 2.2 The 1987 supernova explosion in the Magellan Cloud, as recorded from Earth.

To summarize, there are at least four major initial criteria for the existence of conditions compatible with complex life in a star system. The first is that the star must be relatively young, having been formed in secondary mode from material of high metallicity, spewed out by a supernova. The second is the mass and longevity of the star. If, for example, a star were twice the mass of our Sun, it would last only 800 My before consuming itself, too short a period for Earth-like evolution of life; moreover, it would radiate much life-damaging short-wave UV radiation. A star of 15 solar masses completes its main sequence in only 10 My. On the other hand, if the star is of very low mass it would tend to have many eruptions, again giving off unsteady radiation often inimical to life. The third criterion is that the star should not be one of a binary system which, as noted, would not support planets in orbits favourable to life. The fourth criterion is the star's position in its galaxy. This is discussed below. These are only initial considerations. Life has numerous other requirements for its emergence and evolution.

Galaxies

Some 2.5 Gy after the Big Bang and some 2.3 Gy after the formation of the first stars, gravity began to pull the stars into groups, or galaxies. A galaxy may be formed of anywhere from a few hundred thousand to many hundred billion stars. There are many forms of these galaxies; no two are exactly alike. Our Milky Way has a spiral disc shape (Fig. 2.3) similar to that of our 'close' neighbour the Andromeda galaxy (Fig. 2.4a). It contains some 100–400 G stars. The Sun is situated on the edge of the Orion–Cygnus arm, about 25,000 ly from the galaxy's centre. It revolves around the centre once every 250 My.

Galaxies may be grouped into clusters, from just a few to a few thousand. There are also clusters of clusters, or super clusters, which are the largest formations in the observed universe. Estimates of the number of galaxies in the universe keep changing (upwards). The latest estimate, from Hubble telescope observations, puts their number in excess of 125 billion. Some galaxies have been observed to pass through or collide with others. When this happens we can only imagine the life-disturbing tidal perturbations that occur. Long-term, more or less steady conditions, as required for extended evolution, would be most unlikely.

The nearest galaxy to our Milky Way is a recently discovered relatively small galaxy, which has been named 'Sagittarius-Dwarf'. It is currently only 80,000 ly away from the centre of our galaxy. The Milky Way halo (which contains very old stars and much ionized gas) may extend as far out as hundreds of thousands of light years. This puts Sagittarius-Dwarf as actually already within the Milky Way. It will probably be entirely absorbed into our galaxy in about 1 billion years. Although it has only 1/1000 the mass of the Milky Way, the gravitational and radiation perturbations caused by this collision could have catastrophic consequences for planetary life around any star in its vicinity. Until the discovery of Sagittarius-Dwarf the nearest recognized galaxy was the spiral Andromeda, which lies at a distance of 2×10^6 ly. It too appears to be on a collision course with the Milky Way, a convergence which may begin some 3 Gy from now.

Within what is called the 'local universe' some 80% of the galaxies are less luminous than the Milky Way. Luminosity is positively correlated to metallicity. Metallicity determines whether stars will have the materials required for the formation of terrestrial-type planets. This, at one stroke, puts 80% of the near universe into a category in which life is far less probable.

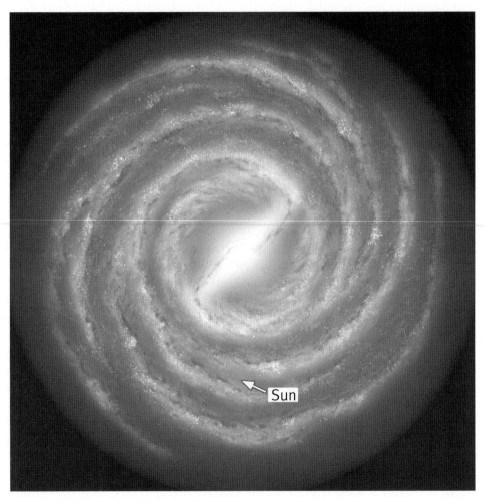

Fig. 2.3 Our galaxy, the Milky Way. Note that, unlike Fig. 2.4, this figure is just an artist's illustration. To photograph the Milky Way like this we would have to place our telescope on another galaxy and look back! Maybe some later edition of this book—in 500,000 or so years from now—will show such a picture! Courtesy of NASA.

Fig. 2.4 Spiral and elliptical galaxies: (a) the spiral Andromeda galaxy. Courtesy of NASA; (b) the elliptical Centaurus galaxy. Courtesy of the ESA.

Galaxies come in many different shapes, from round to elliptical. This is significant when considering the possibility of their having stars with planetary environments capable of supporting life. In elliptical galaxies, such as the Centaurus Galaxy (Fig. 2.4b), the stars tend to have disorderly orbits and may find themselves heading for the centre of the galaxy. There, the density of the other stars would produce an environment of high radiation and gravitational perturbation inimical to life. Moreover, the statistical chance of being in the neighbourhood of a supernova is much higher in the star-dense galactic centre. (On the other hand, star formation and supernovae tend to be less frequent in elliptical galaxies.) Some 80% of the observed galaxies are elliptical, which tends to rule them out as promising homes for life. Taking the reciprocals of the above two 80% factors and multiplying leaves just 4% of galaxies as attractive candidates for harbouring stellar systems with environments suitable for life. However, this is still a *very* large number.

The position and motion of the Sun within the galaxy may have important consequences for life on Earth. It has been suggested that when the Sun crosses the denser part of the arm of the Orion–Cygnus galaxy there may be gravitational perturbations of the Oort cloud of comets. Some of them could affect Earth. Rather surprisingly, these movements of the Sun may also affect the weather on Earth (see chapters 3, 5, and 9).

Planets of other star systems

The astrobiology of Earth has its roots in general astrobiology, the science of life in the universe. In the last two decades we have witnessed the extraordinarily exciting discovery of numerous extra-solar planets (ESPs) orbiting other stars. This has been the result of the development of indirect methods for detecting their presence. The first method is based on the observation of small wobbles in the motion of stars. Such movements are induced by the presence and gravitational pull of orbiting planets. The wobbling can be used to calculate the mass and orbits of the planets. Where the planet passes in a plane between the star and our Solar System, it is possible to detect a small but cyclic reduction in the star's luminosity. This is similar to the transit of the Sun by Venus (itself a rare occurrence), as seen from Earth on 8 June 2004. To date, both methods have been able to detect only very large planets, many times the mass of Earth. By early 2009 some 300 ESPs have been found by these methods. Three planets, two of which are only 10 times the mass of Earth, accompanied by an asteroid belt, have recently been detected orbiting a star at a distance of 42 ly in the Puppis constellation (Lovis *et al.* 2006). A conservative estimate is that at least 10% of stars have planets.

The difficulty which optical telescopes must overcome in the search for ESPs is the very low (reflected) radiation of such bodies. The light reflected by an ESP is outshone by that of its star, which at stellar distances appears in close proximity. The space-borne Hubble telescope (free of the shimmer caused by the Earth's atmosphere) has been able to view some very large extra-solar planets and even detected atoms of oxygen and carbon in the atmosphere of an ESP. As of early 2009, there are advanced plans to build very large computer-stabilized telescopes on Earth and in space for the optical search for Earth-sized ESPs. A very recent discovery, made with a new method of planet detection based on the microlensing effect,[1] has revealed a star system with planets very similar in mass and position to those of Saturn and Jupiter in our own Solar System (Kerr, 2008, reviewing a paper by Goudi *et al.*, 2008). This system is at a distance of 5,000 ly from Earth.

Putting a damper on the enthusiasm engendered by the discovery of ESPs, Mario Livio of the Hubble Space Telescope Institute, writing before the paper by Goudi *et al.* (2008), pointed out that all the ESPs detected so far have a much greater mass than Earth. Moreover they have highly elliptical orbits which would produce unstable radiation and meteorological conditions very difficult for life. This is in contrast to the almost circular orbits of the planets of our Solar System. An elliptical orbit offers the same negative considerations for life as discussed for planets orbiting binary star systems. However, these massive, elliptically orbiting planets are the ones most easily detected by our present technology. An exception to this is a planet orbiting the star HD 28185 in the Eridanus constellation. It was first discovered in 2001. The planet is very large by the standards of our Solar System, with a mass about six times that of Jupiter. However, it is within the star's 'habitable zone', meaning that its surface temperature is calculated to be able to support liquid water. Moreover its orbit is almost circular with a 'year' of 383 ± 2 days.

Stars must have high metallicity to produce planets. At the time of writing no planets have been found orbiting stars with a metallicity less than 40% that of the Sun. On the other hand, if a star's metallicity is very great it may have produced giant planets, many times the size of Jupiter. The presence of such a giant in a planetary system may well perturb the orderly circumnavigation of smaller Earth-like planets.

Comets and asteroids

Stars form from the accretion of mass 'floating' in space. A large enough mass becomes so dense that fusion of hydrogen commences, driving its internal furnace. Smaller bodies do not achieve this required critical mass and remain relatively cool. They tend to circle a star, held by mutual gravitational attraction. The larger of these bodies are the planets. Smaller remnants of the early stellar system are comets, asteroids, and moons (satellites of planets). Although probably common to other stellar systems, we know of asteroids and comets mainly through those in our own Solar System. Planets are the main candidates in the search for places where life may be found, but the

---

[1] As predicted by Einstein and subsequently demonstrated, gravity attracts light waves, producing a lensing effect. This can be used to detect stars and large planets.

emphasis is again on 'may'. Moreover, certain moons of planets in our Solar System may harbour at least primitive life (see Chapter 3). The recent discovery by Lovis *et al.* (2006), noted above, supports the expectation that, in addition to planets, asteroids, comets, and planetary satellites should be common in other high-metallicity star systems. The comets and asteroids in our Solar System are discussed in greater detail in Chapter 4.

A brief survey of some other denizens of the universe

> *Not only is the universe stranger than we imagine, it is stranger than we can imagine.*
> Sir Arthur Eddington (1882–1944)

The universe contains numerous objects, some very bizarre. They may have biological significance, but probably only very indirectly. A few are briefly presented here.

After a supernova explodes outwards it implodes and becomes either a black hole (see below) or what is called a neutron star. Such neutron stars often radiate intensely, but mainly from their magnetic poles. As they also spin rapidly (often in less than a second) they produce beams of periodic, flashing radiation. Hence they are called pulsars or pulsating stars. For some time before they were better understood, their original discoverers, and of course science fiction buffs, believed that they might be celestial lighthouses, set up by superior civilizations to signal their presence. In the absence of a physical explanation the first pulsars discovered were playfully designated as LGM1, −2 ... (for Little Green Men). Sadly, with understanding came disillusionment.

Black holes are among the most studied and bizarre objects in space. A black hole is a remnant of a giant star many times the mass of our Sun. Black holes may vary in size from sub-particle (at present just a theoretical projection) to the order of kilometres. Their mass is so great that gravity does not allow even radiation, such as light, to escape. As the name indicates, we can see nothing but a black hole. Their presence can also be deduced from the great angular velocities attained by nearby stars and the matter falling into them. There are numerous black holes in our galaxy and there seems to be one at its centre. As far as life is concerned the main lesson is—stay away. No planet or star ship could escape the pull of a black hole.

Even more exotic than black holes are wormholes, first predicted in 1935. They are theoretical derivations from Einstein's relativity equations. However, no evidence has yet been obtained of their existence. According to theory, wormholes may connect different parts of space and time, allowing access from one place in the universe to another. If we accept the existence of parallel universes, as predicted by some cosmologists, they may even connect one universe to another. Lack of solid proof of the existence of wormholes and parallel universes has not prevented their appearance in numerous science fiction novels and movies.

Galactic and Circumstellar Habitable Zones

The concept of a galactic habitable zone (GHZ) has recently been proposed and elaborated by a number of astronomers, such as Gonzalez *et al.* (2001b) and Lineweaver *et al.* (2004) with a more popular review by Gonzalez *et al.* (2001a). It is an extension of the older idea of habitable zones (HZ) around stars (Franck *et al.* 2002), now called circumstellar HZ or CHZ. In brief, a GHZ is the region in a galaxy where initial conditions exist for supporting life in a stellar system. The limits of a GHZ are determined primarily by two considerations: metallicity, which allows for the formation of terrestrial planets, and low star density. The latter affords life-bearing planets a greater statistical chance of being far from extreme lethal radiation events. An artist's impression of the solar CHZ within the GHZ of the Milky Way is shown in Fig. 2.5.

The CHZ, as depicted for Earth in Fig. 2.5, is usually delineated as the region around a star in which the surface temperature of a planet with an atmosphere will allow water to exist in the liquid phase for long periods of time (measured in billions of years). The absolute necessity for the presence of liquid water for life was argued in Chapter 1. The CHZ of the Solar System is described in more detail in Chapter 3.

The Milky Way GHZ

As described above, our Milky Way galaxy is spiral shaped (Figs 2.4 and 2.5). Stars within the galaxy tend to rotate in more or less orderly orbits, around the galactic centre.

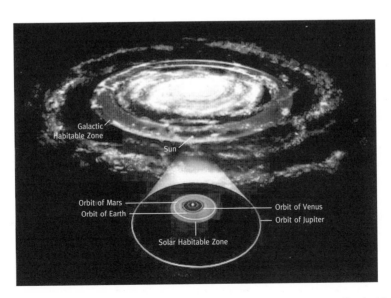

Fig. 2.5 The Milky Way galactic habitable zone (GHZ) and the Sun's circumstellar habitable zone (CHZ). Courtesy of Scientific American.

This is contrary to the transit of stars within galaxies of less circular shape, such as the elliptical Sagittarius-Dwarf. In such galaxies the orbits of stars are more erratic. They may pass through very dense regions where there is a great likelihood of encountering radiation bursts of supernovae or intense gravitational perturbations, both inimical to life. Even with a galaxy such as the Milky Way, whose shape allows relatively constant conditions for its star systems, large regions of the galaxy must be ruled out for life. This leaves the *relatively* small GHZ.

The Milky Way has four main regions: the central bulge, a thick disc, a relatively thin disc, and an outer halo. The central bulge has a very high density of stars, some with the high metallicity required for the formation of terrestrial planets. However, as in the centre of elliptical galaxies, any star system in this region would be likely to encounter extremes of radiation and gravitational perturbations.

The very large Milky Way halo and also the thick disc region contain mainly older stars of low metallicity. Their planets would not have the materials required for life. The main region of the Milky Way which may be amenable to life is the thin disc where, not surprisingly, our Sun is to be found. Our Sun's metallicity (as noted above, itself just a few per cent of the total mass) may be used as a baseline. Stars having between 60% and 200% of this metallicity exist mainly in a region circling the centre of the galaxy at a distance of approximately 15,000–38,000 ly. The GHZ is therefore considered to be in this ring (the tyre-shaped 'galactic habitable zone' depicted in Fig. 2.5). It contains only some 20% of the total stars in the galaxy.

By no means all stars in the GHZ have the other characteristics required for life. A first limitation is that there are two types of metallicity which may produce planets. The first contains mainly cobalt, iron, and nickel which are produced at the end of the life of a white dwarf star. The second type also contains calcium, magnesium, oxygen, and the other elements used by life. As discussed above, they are produced by implosion of a massive star, followed by its explosion (as a supernova). Our Sun is rather exceptional in being both long-lived and having a 40% greater metallicity than most stars of the same age, with a metallicity belonging to the second group.

There are two main conclusions from this biological review of the universe. The first is that the universe is large, far beyond our daily experience and almost beyond human comprehension. The second is that only in a very small percentage of the vast regions of the universe do the primary conditions required for life exist. Even so, when we multiply these two factors there are still millions of stars, even within our own Milky Way galaxy, where life could develop if numerous other requirements are also met. The chances of contacting extraterrestrial intelligent life and of navigating and colonizing the galaxy are discussed in Chapter 10.

Summary

(As in the main text, where 'life' is mentioned it is limited to life as we know it on Earth.)

- The universe is *very* large. The diameter of our Milky Way galaxy, is 100,000 ly. The distance to the nearest star is 4.3 ly.
- The Milky Way has some 100–400 G stars and the number of galaxies in the universe is more than 125 G.
- Time is understood with reference to before and after events. It is character-ized by having only one direction. It is apparently irreversible. Life depends on a wide time range of reactions, from those that take fractions of a second (e.g. in photochemistry) to processes which take billions of years, such as evolution.
- In relation to time and distances in the universe the human life span is vanishingly short. This is a major limitation for the exploration of the cosmos.
- Electromagnetic radiation, such as visible light, travels at the fastest possible speed (almost 300,000 km s^{-1}), but this is relatively low in relation to the size of the universe.
- Only a small part of the physical composition of the universe can be used by life. Elements essential for life, including those heavier than hydrogen, such as carbon and iron, constitute about 0.03% of the total.
- Most locations in the universe are not suitable for life:
  - 80% of the galaxies nearest to the Milky Way have low luminosity, which is correlated with low metallicity, meaning a low content of elements essential for life;
  - about half of the galaxies have a non-spherical shape: stars in such galaxies may have irregular orbits that bring them into regions with a high frequency of supernovae;
  - the galactic habitable zone (GHZ) is the region in a galaxy where the type, density and orbits of the stars show a potential for planetary life—the GHZ of the Milky Way is about 20% of the total;
  - about half the stars are binary: their planets would not have orbits suitable for life;
  - only a small proportion of stars have the mass and longevity required for the evolution of life on their planets;
  - within a solar system only a small region, the circumstellar habitable zone (CHZ), can provide planets with environments suitable for life; especially surface temperatures which allow water to be in the liquid phase.
- So far (in early 2009) some 300 extra-solar planets have been detected. None has been Earth like. However, this may be a result of the limitations of the detection methods, which favour planets with very large mass and highly elliptical orbits.
- The universe has numerous bodies of great interest to astronomers, and to anyone with an interest in the cosmos, such as black holes, wormholes (if they exist), pulsars, etc. However, their significance to life, if any, is unclear.

3

The Solar System and life on Earth: I

Main components of the Solar System

In the previous chapter the entire universe was considered from the point of view of biology. However, the only place in the universe where, as of early 2009, we *know* that life exists is within our own Solar System. Within this system we are only sure of the existence of life on Earth, and on Earth only in the biosphere, a thin layer just a few kilometres thick. Ask any estate agent and you will be told that the three most important considerations for evaluating property are location, location, and location. The same goes for life in the universe. As discussed in Chapter 2, a favourable location is the central theme of the concept of the GHZ and the CHZ (Gonzalez *et al.* 2001a,b). The Sun is within the GHZ of the Milky Way galaxy and the Earth is within the CHZ of the Solar System. Parameters which define the CHZ are given below.

The main components of the Solar System are its central star, the Sun, planets and their satellites (or moons), comets, and asteroids. To this must be added the all-important radiations: short-wave EM radiation and the particulate solar wind and cosmic radiation (the latter from outside the Solar System).

In the last decades numerous automated probes have been sent by NASA and the European Space Agency (ESA) to most of the solar planets and to many comets, asteroids, and planetary satellites (moons). They have enormously increased our knowledge and understanding of the Solar System. One of the most extensive collections of information on the Solar System can be found in the encyclopaedia edited by Weissman *et al.* (1999). A shorter, more life-focused coverage can be found in Jones (2004). Research on the Solar System is advancing at such a rapid rate that books and reviews need constant updating, including of course the present book. In accordance with the central theme emphasis is placed here on those factors in the Solar System which affect life on Earth.

Our star, the Sun

Early humans and later civilizations recognized, and even worshipped, the Sun as the supporter of all life on Earth. Although this is all too understandable, today we know that the greater part of life on Earth, as measured on a mass basis, actually exists as microbes below the surface, entirely independently of the Sun. Moreover, in the vicinity of deep sea vents there are quite advanced ecosystems with food chains that start with chemo-autotrophic bacteria and reach quite advanced animal forms. They exist without bene-fit of solar-driven photosynthesis (these and other Earth extremophiles are discussed in Chapter 8). The existence on Earth of life forms independent of solar radiation is of considerable significance in our search for regions of other planets inside or outside the Solar System that are favourable for life. Today most advanced life forms on Earth are directly or ultimately dependent on food and energy derived from photosynthesis.

The detailed functioning of the Sun and its formation together with that of the entire Solar System are outside the scope of this book. However, four factors or events in this process must be mentioned as they were and are of major significance to life on Earth:

- The first factor is that the Solar System was formed from the debris of a supernova explosion. As such it contains a small but highly significant 'metallic' content. As explained in Chapter 2, 'metallic' in this sense means elements heavier than hydrogen and helium, such as those essential to life—carbon, sulphur, iron, etc. These elements are the materials (in addition to hydrogen and helium) from which the planets and other extra-solar bodies of the Solar System are formed. The Sun itself is composed of hydrogen (71% by mass) helium (27.1%), and just 1.9% metallic elements.
- The second factor is that 99.86% of the mass of the Solar System is within the Sun, while 90% of the remaining mass is contained in just two of the planets, Jupiter and Saturn.
- The third major consideration for life in the Solar System is the time factor. As dis-cussed in Chapter 2, the Sun is a main sequence star, with a projected lifetime of some 10 billion years. It was formed about 5 Gy BP and will probably survive for another 4–5 Gy before its hydrogen is exhausted by nuclear fusion. The Earth was formed 4.55 billion years ago. Life appeared on Earth about 1 Gy after its formation and has evolved ever since without ever suffering complete extinction (i.e. evolution never stopped, to be followed by a re-emergence of life). On this time scale humans appeared only very, very recently. In other words it took more than a quarter of the entire lifetime of the universe since the Big Bang (13.7 billion years ago) for intelli-gent life to evolve within the Solar System. Judging by life on Earth (and it should again be remembered that we have only this one case to go by) star systems that are not so long-lived would be unsuitable for the evolution of advanced, intelligent life forms.
- The fourth factor is the mass and distance of the planets from the Sun. Their con-stantly changing spatial relationships and mutual gravitational attraction affect each other and, most importantly for us, affect the motions of Earth in its passage around the Sun. These perturbations are small but sufficient to modify the climate of the Earth's biosphere, challenging life with new conditions. When this happens life must survive under stress—adapt or die.

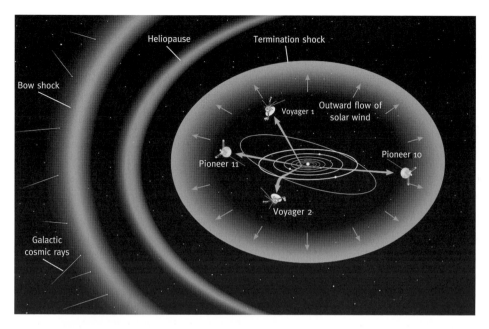

Fig. 3.1 The Heliosphere; showing the present position of humanities furthest celestial probes. Courtesy of NASA.

The heliosphere and solar wind

Around the Sun, extending to approximately 100 AU, is the heliosphere, the region of the solar wind (Figs 3.1 and 3.2). The nature and importance to life of this 'wind' is discussed in Chapter 4. The exact geometry of the heliosphere is not known. It is thought to extend some 100 AU in advance of the moving Sun and much further in its wake. As shown, the automated probes Pioneer 10 and 11 and Voyager 1 and 2, although travelling at great speed (about 3 AU per year) and launched from Earth decades ago, are still within the heliosphere. They must still travel a thousand times further before they reach and pass the Oort cloud[1] on their way out of the Solar System. This is a very humbling consideration in humankind's desire to explore the universe.

The Sun's short-wave EM radiation

The star classification which puts the Sun in the type-G group is based mainly on the star's apparent colour spectrum, which is determined by surface temperature. Type-G stars emit yellowish radiation, as their surface temperatures are between 5000 and

---

[1] The Oort cloud is a sphere of comets which is thought to exist around the Sun at a distance of $\sim 10^4$ AU (see Chapter 4).

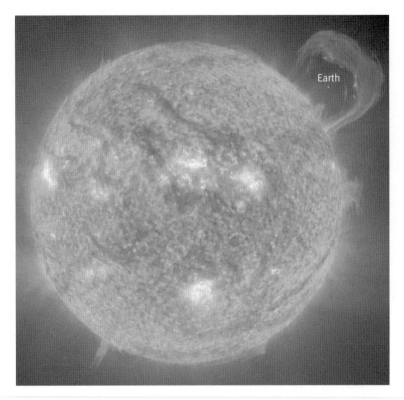

Fig. 3.2 Our star the Sun. The surface temperature is 5,777 K, which gives the familiar short-wave radiation. Note, however, the corona which resembles an atmosphere. The corona is a very rare plasma with a temperature of many million kelvin. The outer layers of the corona move into space, becoming the 'solar wind' (see text). The white areas are called faculae. They are regions of high radiation and are most prominent at times of increased sunspots, which themselves are regions of reduced radiation. Courtesy of NASA.

6,000 K. The Sun's surface temperature is about 5,770 K, which gives us the familiar EM short-wave solar spectrum between 300 and 3,000 nm (Fig. 3.3). Part of this radiation, in the 400–700 nm band, can be seen by the human eye. 'Yellow' is what we call the colour of the waveband between about 500–600 nm.

The term 'short wave' is purely arbitrary. The lower the surface temperature of a body the longer the wavelength of its EM radiation and the weaker the energy in its photons. Radiation coming from cooler bodies, such as the Earth (~300 K), in the range of 5–100 μm, is termed 'long wave'. Our knowledge of the Sun's surface temperature is of course obtained indirectly from measurements of its radiation.[2]

---

[2] Derived from the Stefan–Boltzmann equation $E = \sigma T^4$, where E is the radiation flux, in W m^{-2}, T is the temperature of the radiating body (in K), and σ is the Stefan–Boltzmann constant (5.67×10^{-8} W m^{-2} K^{-4}).

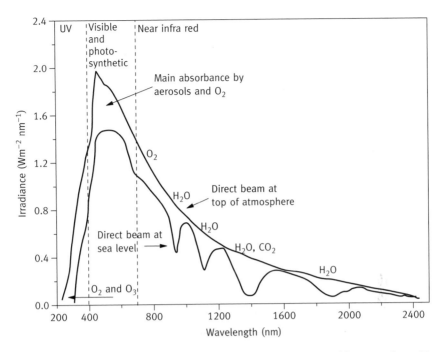

Fig. 3.3 The short-wave solar spectrum and its attenuation by the Earth's atmosphere. Note the absorption of specific parts of the solar spectrum, mainly by water vapour, oxygen, ozone and CO_2.

The Sun is a dynamic body surrounded by a constantly changing corona (Fig. 3.2). The corona is a very rare, very hot plasma of ionized particles. Its thickness is variable. It can extend to a distance of millions of kilometres. Its manifestation varies with the 11-year sunspot cycle (see below). The outer part of the corona radiates into space, producing the solar wind (Fig. 3.1). As discussed in Chapter 4, the solar wind together with the particulate radiation reaching Earth from outside the Solar System (cosmic radiation) affects cloud formation and hence the weather on Earth.

The solar constant is defined as the amount of incoming EM radiation from the Sun (solar radiation) per unit area reaching the Earth. It has been determined by satellites orbiting above the atmosphere to be 1,366 W m^{-2}. While on a day-to-day basis this radiation is constant to within 0.1%, it varies to within ~2% on an 11-year cycle. This cycle is correlated with the appearance of relatively dark sunspots. The spots themselves are regions of reduced radiation, but they are accompanied by areas called 'faculae', where radiation is greater. Short-wave radiation is actually greatest at times of maximum sunspot activity. The cycle is itself quite constant, although the appearance and intensity of the sunspots and faculae are not. Such small variations in short-wave radiation may persist for tens of years, contributing to periods of reduced or increased Earth temperature. For example, from 1645 to 1715 there was a period of very low sunspot activity, the so-called 'Maunder minimum', followed by the 'Dalton minimum' between 1720 and

1820. At this time there was a considerable reduction in measured air temperatures, at least as recorded in Europe and North America. This was the so-called Little Ice Age. By way of illustration, during this period the River Thames in London would freeze over every winter (as vividly described by Charles Dickens), something which has not happened in the last century and a half. However, the connection between sunspots and weather is just a correlation, and there are many system feedbacks in the relationship between solar radiation and climate (Rind 2002).

Within the lifetime of Earth (4.55 billion years) the radiation of the Sun has steadily increased. Since the beginning of the evolution of life on Earth (3.7 million years ago) the increase has been of the order of 20%. Compared with the variations in intensity measured on a day-to-day and an 11-year cycle basis, this is an enormous change (Sagan and Chyba 1997). It resulted from the steady expansion of the Sun while its surface temperature, and hence radiation spectrum and radiation intensity per unit surface area, remained more or less constant. The early Sun was about 10% smaller in size than it is today. A major conundrum in the astrobiology of Earth is how the surface temperature of Earth has remained relatively constant throughout the last 3.7 million years. We know this to be true, despite periodic ice ages, as life, and hence by necessity liquid water, has always been present at least somewhere on Earth. The factors responsible for this rather amazing stability are discussed in Chapter 7. This problem is often called the 'faint early Sun paradox'.

The solar radiation spectrum above and below the Earth's atmosphere is depicted in Fig. 3.3. As shown, the solar radiation which reaches the surface of Earth is very different from that incident on the top of the atmosphere. These modifications of the intensity and spectrum, resulting from the presence of our atmosphere, is essential for life. Note the 'our': other solar planets have atmospheres with different densities and composition, and radiation reaching their surface is never as clement.

As seen in Fig. 3.3 the atmosphere attenuates the total short-wave radiation by about 50%. Even at low latitudes, at noon on a cloudless summer day, solar radiation at sea level rarely exceeds 800 W m^{-2}. Apart from absorption and reflection by dust, aerosols, and clouds, the atmosphere absorbs radiation at certain wavelengths. This is due to the presence of specific molecules such as oxygen, water vapour, and ozone. As a result the atmosphere absorbs heat which is then dissipated by long-wave radiation both upwards into space and downwards toward the Earth.

The solar spectrum above the atmosphere contains about 3% UV (below 350 nm). Most of this radiation is absorbed in the atmosphere. The photons of this waveband are highly energetic, and if absorbed by biological molecules such as proteins or DNA may cause damaging ionization and structural breakdown, which could be mutagenic, carcinogenic, or even directly and rapidly lethal. Part of the longer UV penetrating the atmosphere may be beneficial. For example, radiation between 300 and 400 nm induces colour formation in some leaves and fruit and is used for vision by insects and birds (see below).

Atmospheric oxygen (O_2) absorbs UV and forms ozone (O_3). Ozone is an even more potent UV absorber. As discussed in Chapter 9, certain chemicals produced by humans can prevent the conversion of O_2 to O_3, reducing the UV-protective effect of the atmosphere.

The Sun's radiation supports life on Earth in four ways: it provides heat, energy for metabolism, signals for various physiological and morphogenetic processes, and enables vision.

Solar radiation and the temperature balance of the Earth's surface

As noted above, in relation to the faint early Sun paradox, life has left a record of uninterrupted presence and evolution on Earth throughout the last 3.7 Gy. As life is entirely dependent on liquid water, the temperature on the surface of Earth must have remained between 0 and 100°C, at least somewhere. Moreover, most advanced organisms prefer temperatures of ~25 ± 10°C. It is important to remember that life 'prefers' such temperatures because that is what was available.

Most of the short-wave radiation which penetrates the atmosphere is absorbed by the oceans, land, flora, and fauna, and ends up as heat. All bodies at temperatures above absolute zero (0 K or −273°C) radiate. Because Earth's average surface temperature is 288 K (15°C), it radiates in the long wave (between 5 and 25 μm). Part of this long-wave radiation is absorbed by the atmosphere and part penetrates the atmosphere and is lost to space (Fig. 3.4). What is not shown in Fig. 3.4 is a small flux of heat coming from the hot centre of the Earth. Although its contribution to the surface temperature is negligible, it is of importance to the wealth of microbes which live below the surface.

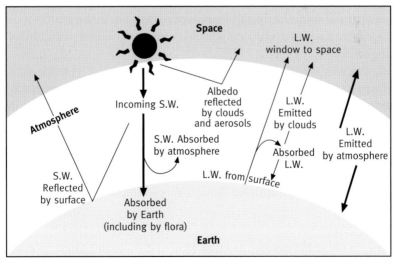

S.W. – Short wave radiation (<3,000 nm)
L.W. – Long wave radiation (>50,000 nm)

Fig. 3.4 The Earth's radiation balance.

It must always be borne in mind that in the course of evolution life adapted to the conditions available. First and foremost was the presence of liquid water, a ubiquitous molecule on Earth but one which has exceptional physicochemical characteristics (see Chapter 1). Life also adapted to the temperatures found in the various geographical regions on land and in the rivers, lakes, and oceans.

As seen in Fig. 3.4, there are many factors contributing to the Earth's radiation balance, which together determine the equilibrium surface temperature. These include the incoming short-wave radiation and its reflection (albedo) by clouds, snow, ice, atmospheric dust, and aerosols. The net short-wave radiation is balanced by the net outgoing long-wave radiation from the Earth's surface (oceans, land, and atmosphere). Some wavebands of the outgoing long-wave radiation are absorbed by specific molecules in the atmosphere. These include water vapour, CO_2, methane, and ozone. As a result these molecules heat up and radiate in the long wave: upwards to space and downwards, back to the Earth. The greater the atmospheric absorption of long-wave radiation the higher the surface temperature. With analogy to agriculture, this is commonly called the 'greenhouse effect'.[3] Humanity is currently affecting this balance by raising the atmospheric content of these 'greenhouse gases'. To what extent this anthropic effect is responsible for 'climate change' is unclear (Shaviv and Veizer 2003; see Chapter 9).

Solar radiation and photosynthesis

The evolution of photosynthesis and its effects on the biosphere are discussed in Chapter 6. The major contribution of photosynthesis to life on Earth is that it introduces external radiant energy into the biosphere, an otherwise almost closed, thermodynamic system. ('Almost closed', as some energy in the biosphere comes from the hot core of the Earth and from the decay of radioactive isotopes in the Earth's mantle.) This enables the local reduction of entropy, which is the most prominent feature of growing, reproducing life forms.

Although the first stage of photon absorption is remarkably efficient, approaching 99% (Engel *et al.* 2007), photosynthesis uses very little of the short-wave radiation reaching Earth. Most photosynthetic pigments absorb and utilize radiation only in the 400–720 nm waveband, which contains about 48% of the total available short-wave energy. The reason for this is that the basis of photosynthesis is the splitting of water molecules and the photons of radiation at wavelengths above 700 nm do not contain sufficient energy to carry this out. Some early bacteria and algae were able to utilize radiation at wavelengths a little above 700 nm by splitting other molecules, such as H_2S,

---

[3] The analogy is useful but not very accurate. Most of the warming inside agricultural greenhouses is a result of the prevention of sensible heat exchange with colder outside air. The return flux of long-wave radiation from the greenhouse roof (which absorbs some of the outgoing long-wave radiation) is only a secondary contributing factor.

Such molecules, whose atoms are held together by lower bond energies, are now relatively rare. Today, representatives of these bacteria are only a minor component of the biosphere. As the energy of photons at wavelengths shorter than 400 nm can be destructive to biological molecules they are unsuitable for photosynthesis.

Although photosynthesis *can* utilize 48% of the available solar energy, in practice it uses much less. The basic theoretical efficiency of photosynthesis (as derived from measurements of the photons absorbed per molecule of oxygen released) is about 12%. Eight photons are used for every molecule of CO_2 fixed. However, absorption of radiation by photochemical pigment molecules is limited. Moreover, the frequent lack of water and mineral nutrients and competition for light and other resources reduce the actual radiation-use efficiency of plants in the field. This latter is defined as the ratio of the chemical energy bound in the plants to the incident short-wave energy flux, per unit ground area. Even under optimized agricultural conditions it is usually below 4%.

Why photosynthesis and plants did not evolve to utilize more of the available energy is probably connected to the environment in which life and then photosynthesis appeared. On dry land, a lack of fresh (non-saline) water was, and still is, the factor most likely to limit photosynthesis. Furthermore, solar energy is often not the limiting resource in photosynthesis, metabolism, growth, and reproduction. Consequently, there was little environmental drive for the evolution of a process that is more conserving of photon energy.

Short-wave solar radiation, signals, and vision

Life evolved on Earth in an environment of short-wave solar radiation which lasts for about half of every 24-hour day (with day length depending on the geographical latitude and season of the year). To state this seems so trivial, yet it is one of the most significant astrobiological factors affecting life. We tend to take for granted the cessation of many of life's activities during dark periods of our 24-hour day. It did not have to be so. Not all solar planets rotate with periods close to that of Earth or revolve around the Sun with a more or less constant axial inclination, a behaviour which produces recurring seasons. Venus, for example, always keeps the same face pointing to the Sun. Life on Earth has adapted to the dangers of daily dark periods and sometimes even utilizes them. Moreover, life has evolved mechanisms which cope with changes in day length and climate. Take, for example, jet lag. We suffer from the rapid transit through time zones (1 hour for each 15° of longitude), but after about 1 day for each hour of change our body's internal clock resets.

An example of the adaptation to Earth's diurnal rotation can be found in plant metabolism. Plants can produce food (by way of photosynthesis) only in daylight hours, but expend it, via respiration, throughout the day. Consequently, the anabolic and catabolic processes must be, and are, balanced in a way which allows growth. Many plants will not thrive if exposed to 24 hours of continuous light, if their respiration and metabolic processes do not increase proportionally.

Sleep is an adaptation of mammals, birds, and many fish and reptile species to diurnal dark periods. Sleep reduces the overall energy expenditure during the period of the day (or season, in the case of hibernating animals) during which energy (food) gathering is minimal. Moreover, advanced brains have evolved to exploit these periods of sleep for 'housekeeping'. Although the mechanism is poorly understood, we are all aware of the mental confusion induced by sleep deprivation.

Shortly after life first learnt to convert solar radiant energy to chemical energy which can be used for metabolism (by way of photosynthesis) it also evolved mechanisms for utilizing light for long-distance sensing of the environment. Among the first such evolutionary adaptations which enabled the use of short-wave radiation for long-distance sensing, was the 'eye spot' of flagellate, unicellular, photosynthetic algae. An example is phototaxis shown by species of *Chlamydomonas* (Kateriya *et al.* 2004). Their flagella propel the alga either towards or away from a light source; towards the source when the intensity is below the optimum for photosynthesis, and away from the source when the intensity would be so high as to cause damage. Their 'eye' contains some four different forms of rhodopsin, a pigment still employed for vision by advanced animal life. These pigments and their associated proteins sense the intensity and direction of radiation and send an appropriate electrochemical signal to the flagella. This is a remarkable system for a unicellular life form which evolved on Earth more than 2 billion years ago. It has all the elements of signal detection, measurement of direction and intensity, decision-making, and transfer of information to an action executive (the flagella). We tend to associate such sophistication with more advanced life forms, such as insects and animals which have advanced vision systems encompassing coordination between highly specialized light detectors (eyes), neural networks, and brains.

Primitive life forms often use radiation for vision and sensing at intensities as low as, or sometimes much lower than, those now used by humans and other advanced animals. Take, for example, the fungus *Phycomyces blakesleeanus*. It has single-celled sporangium-bearing stalks which bend towards or away from light; a response called positive or negative phototropism. Fungi evolved even before photosynthesis. Their 'vision' is more sensitive to light than the dark-adapted human eye. Such fungi and many other plants with phototropic responses use flavin pigments which absorb in near-UV wavelengths around 370 nm. A negative response is possibly an advantage in avoiding over exposure to UV radiation. The advantage of a positive response to light is clear for photosynthetic algae or higher plants, but not for the non-photosynthesizing fungi. Phototropism has long been studied. In fact this was a main research subject of Charles Darwin, better known for his other interests, in the 19th century.

Plants use light signals for very many functions other than phototropism. They detect very low light intensities of certain specific wavelengths. Among many other things this can be used for deducing the length of dark periods and hence the season of the year. This information may then signal the induction or suppression of seed germination, leafing, flowering, shoot elongation, food storage versus leaf formation, etc.

Consider just one example of sensitivity to radiation signals. Plant leaves are often seen to grow in a pattern which avoids mutual shading. The mechanism behind this phenomenon is that leaves selectively absorb the red parts of the solar spectrum. This results in the light transmitted through the leaf having a higher percentage of long-wave radiation (>700 nm). New, nearby leaves can sense such longer-wave radiation and respond by suppressing growth in that direction.

The pigment responsible for very many of these responses is phytochrome, a chromo-protein. Phytochrome (P) acts as a biological switch. Basically it is a molecule which is formed in darkness. On exposure to red light (R), of wavelength 660 nm, it takes on a biologically active form (called P_{FR}) which can induce morphogenetic reactions. Exposure of plants containing the P_{FR} form to far-red radiation (FR) at 730 nm reduces the phytochrome to the biologically inactive, reduced form (P_R).

The R/FR ratio of solar radiation which reaches the surface of Earth changes at twilight and dawn. At these times it contains a higher fraction of FR radiation. This is a result of atmospheric refraction when the Sun's angle is low. Refraction, which diverts part of the solar radiation back to space, is stronger at the shorter wavelengths. Plants are able to sense the length of the night between these two periods, producing responses called photoperiodicity. Such responses vary in different species and varieties. In respect to flowering response, plants are termed long-day, short-day, or indeterminate. Species originating in the tropical latitudes, where there is little seasonal change in day length, are usually indeterminate. The switching effect of phytochrome is so incisive that periods of artificial night-time illumination, sometimes as short as minutes, are used by horticulturalists to induce or suppress flowering in certain market crops, such as chrysanthemums.

How vision in higher life forms developed from the above is not a primary subject of the astrobiology of Earth. It is, however, relevant to note that evolution has not placed humans and other primates at the apex of visual acuity. Some animals are able to utilize a wider range of the solar spectrum which reaches the Earth's surface. Birds, lizards, turtles, and some fish have four types of cone cells in their eyes. These cone cells enable vision in the near-UV (300–400 nm) range. Many early mammals lost two of the four cone cell types, restricting the spectrum of their vision. They were then left with vision which is most sensitive at the yellow–green 530–560 nm wavelengths. Later in evolution, some mammals, including humans, regained one of the cone types which has maximum sensitivity at 424 nm, but they still have very limited near-UV vision. The four cone types are advantageous at very low light intensities and for distinguishing colours in plants and insects which are quite beyond primate, including human, vision.

The concept of a Circumstellar Habitable Zone (CHZ)

Planets such as the Earth are large bodies orbiting stars; they do not support nuclear reactions other than the disintegration of remnant atoms of radioactive elements. Being relatively cool, planets do not radiate in the short wave other than what they reflect from their star (their albedo). Planets and their satellite moons are considered to be prime candidates as sites where life could have evolved.

Where in a star system would there be a region suitable for planets with conditions amenable to life? This is similar to the question of life-suitable locations of stars in a galaxy, which led to the development of the concept of a GHZ (see Chapter 2). Within a star system, such as that of our Sun, this is called the CHZ. (Note that some authors refer to the 'continuously habitable zone' rather than the 'circumstellar habitable zone'. They emphasize the importance of long-term steady conditions as required for the evolution of advanced life. Here again, they take the Earth, with its only known incidence of life, as a model.)

The basic requirement for a CHZ is that surface temperatures on planets within this zone must be such that water can exist in its liquid form. Too far from the star and any water present would freeze; too close and it would boil away. Additional considerations for a CHZ are the size and radiation of the system's star and the size, mass, and atmosphere-holding abilities of its planets. Basic equations for a CHZ were developed by Kasting *et al.* (1993). Their model includes planetary atmospheres and their effects on surface temperatures. Special attention was given to what is known of the present state and histories of the terrestrial planets, Mercury, Earth, Mars, and Venus.[4] Their model gives estimates of the CHZ boundaries of the Solar System: 0.75–0.95 AU for the proximal boundary and 1.37–1.90 AU for the outermost boundary. Earth is at 1 AU—recall that an AU is defined as the average distance between the Sun and the Earth.

Franck *et al.* (2000) further developed the CHZ model. For the Solar System they took into consideration the interactions between increasing solar luminosity, carbon cycling as driven by photosynthesis, and tectonics (see Chapter 4). Earth, the only planet in the Solar System which has tectonic activity, was very close to the ideal distance from the Sun, which they calculated to be 1.08 AU. Their model suggests that Mars, at 1.5 AU, would have been habitable up to about 500 million years ago. Venus, at 0.72 AU, would always have been too close to the Sun for the evolution on its surface of conditions suitable for life.

Present CHZ models do not include secondary regions where life not based on photosynthesis could exist. As considered below, certain satellites (moons) of planets of our Solar System which are located far outside the CHZ may have conditions for the development of at least primitive life (Reynolds *et al.* 1987). Radiation from a star is not the only possible source of heat and energy which can satisfy the requirements of life. This is the case for much of the microbial life on Earth, which lives below the surface (Gould 1999; Chapter 8).

---

[4] The 'terrestrial planets' are characterized by having iron cores, a composition dominated by silicates, and solid surfaces. This differentiates them from the gas giants Jupiter and Saturn. The outer planets Neptune and Uranus are formed of ice and rock.

Planets of the Solar System and life on Earth

The most exceptional planet in the Solar System is of course Earth. Its very special characteristics are what have enabled the appearance and evolution of life. However, conditions on Earth are also influenced by the other planets. They affect life on Earth in two ways. First, as noted above and in greater detail in Chapter 4, their gravitational attraction may perturb the motion of the Earth around the Sun. The Earth rotates on its axis and circumnavigates the Sun in a complex pattern. The geometry of these movements determines the flux of radiation reaching the different regions of Earth, and the seasons. Any change in these movements affects Earth's climate. This contributed in the past to Earth having suffered through periods of deep cold (ice ages) and interim warmer periods (Chapter 7). Note the word 'contributed'—there were many other factors.

Contrariwise, the relatively close presence of the massive planet Jupiter has a stabilizing effect on the movements of Earth. Moreover, it may reduce the frequency of comet and asteroid impacts, which can be and have been catastrophic for life, although this long-accepted effect has recently been challenged (see chapters 5 and 7). Models of the formation of the Solar System suggest that the presence of such a massive planet as Jupiter conserved the Sun's angular momentum and stabilized the entire system, especially the motions of the inner, terrestrial planets.

The second effect of other planets on Earth is as a possible source of early life. This latter idea is called 'Panspermia' and is discussed in Chapter 6 in relation to the origin of life. In brief, this theory suggests that 'seeds' of life, in the form of meteor-borne microbes or other packages of biological information, may have reached Earth from other bodies in the Solar System, or even from extra-Solar System bodies (see Fig. 1.4 for example). This concept begs the question of where else in the Solar System might there be conditions conducive to life which pre-dated that on Earth?

Our print and electronic news media are full of the immediate effects of the planets on our daily life, as expounded by astrologers. This after 5000 years of speculation during which time humanity should have learnt better. After dismissing such superstition (see Chapter 10, in relation to the history of astronomy and astrobiology), we recognize the above two possible effects of the planets on life on Earth. The mutual gravitational effect is certain, but hard to quantify. The panspermia theory is possibly correct but is unproven and far more speculative.

From the discovery of Pluto in 1930 up until 2006 nine planets were recognized in the Solar System (Fig. 3.5). However, astronomers have now demoted Pluto to the class of 'dwarf planets'. This was because of the irregular plane of its orbit, which is at a steep angle to the ecliptic, the plane of Earth and the other planets in their circumnavigation of the Sun. In addition, its small size and probable origin make it very different from the other eight planets. At present there are two other recognized dwarf planets: Ceres, which lies within the asteroid belt (see Fig. 4.1), and Eris, which is just outside the Kuiper belt (see Fig. 4.2). Eris has a diameter of 2,400 km, which makes it a little larger than

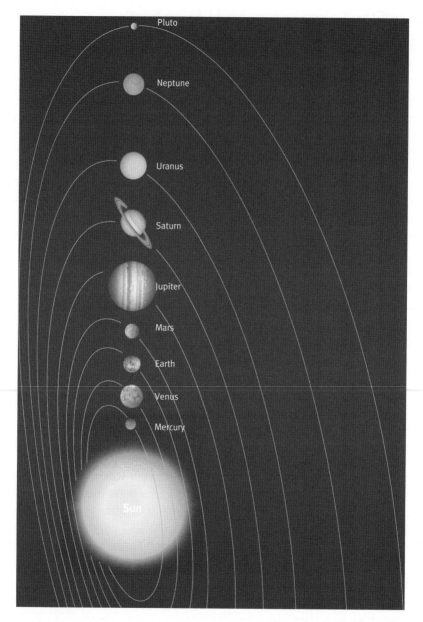

Fig. 3.5 Planets of the Solar System. Note: as discussed in the text, Pluto is now defined as a dwarf planet. Not to scale. Courtesy of NASA.

Pluto. For a full description of the planets readers are referred to Weissman *et al.* (1999) and Beatty *et al.* (1999). The latest information on the field of planetary research, which is advancing very rapidly, can be found on the websites of NASA and the ESA (see References and Resources).

Some extraterrestrial sites in the Solar System which may support life

Our human yearning not to be alone makes it of special interest to seek out other places where life could possibly exist. The search is for life, any life, be it simple or advanced, past or extant. Unlike other star systems which may harbour even advanced, intelligent life, the Solar System is accessible to close human exploration, at least by unpiloted, automated probes. As discussed in Chapter 10, life in other Solar Systems can only be studied at long distance by the analysis of radiation reaching Earth.

There are a number of sites in the Solar System which have potentially life-supporting physicochemical conditions. Some are well outside the CHZ, as commonly defined. The characteristics of extremophiles on Earth make it possible that similar life forms evolved in these special niches. The panspermia hypothesis suggests that any microbial life forms engendered in these niches may have been carried to Earth (the reverse direction is also possible!).

The terrestrial planets—Mercury, Venus, and Mars in addition to Earth (see above)—are relatively close to one another and to the Sun. When looking for extraterrestrial life they are of course the first candidates for study.

Mercury

Mercury is the smallest of the terrestrial planets and nearest to the Sun (Fig. 3.5). At 0.39 AU it is far too close to the Sun to be within the CHZ. It is small, dense, and rocky, with almost no atmosphere. It revolves slowly, with days which are about 60 times longer than those of Earth. This gives it a huge range of surface temperatures, from 427°C during the day to −183°C, at night. However, there are shaded craters near its polar regions where radar studies suggest the presence of water ice. Life on Earth has surprised us so often with its presence in unpropitious sites that perhaps even here microbial life cannot be entirely ruled out, even though, without an atmosphere, Earth-type microbes would soon dehydrate.

Mars

Mars is the nearest planet to Earth and has always attracted speculation as to the presence of life, even intelligent life (see Fig. 10.2). Compared with Earth it is far less suitable for the evolution of advanced life, which requires very long periods of relatively stable conditions, something specifically absent on Mars. In its present conformation it is very unlikely to have even the most elementary biosphere (defined as a life-supporting niche with extended longevity). There may, however, be subsurface ecological niches which can support microbes. Such life forms may have survived since the time when water and a thicker atmosphere almost certainly existed on the Martian surface. A comparison of some of the more important characteristics of Mars and Earth relevant for life is given in Table 3.1. Leovy (2001) has reviewed the present weather and climate of Mars.

Table 3.1 A comparison of some of the characteristics of Mars and Earth relevant to the development and evolution of life

	Earth	Mars
Distance from the Sun	1 AU	1.5 AU
Solar constant (average)	1366 W m^{-2}	600 W m^{-2}
Eccentricity	±3.3%	±19%
Moons	One, large	Two, small
Length of day	24 h	24.6 h
Length of year	365.25 days	678 Earth days
Gravity (acceleration of free fall, g)	1 g (9.81 m s^{-2})	0.38 g
Main constituents of the atmosphere	78% nitrogen, 21% oxygen	95% carbon dioxide
Atmospheric pressure at sea level	1 bar	0.007 bar[a]
Surface temperature	−90 to +58°C	−170 to −8°C
Liquid water on surface	75% of surface	None
Water ice	Covers poles	Possibly some subsurface permafrost

[a] Pressure at the lowest average land level. There are no oceans on Mars.

As seen in Table 3.1, although Earth (at 1.0 AU) and Mars (at 1.5 AU) are relatively close, both within the calculated CHZ (the inner and outer limits of the CHZ being 0.75–0.95 AU and 1.37–1.90 AU, respectively) and are in many ways similar, there are very significant differences between their physical and chemical environments. Today, conditions on Mars are such that there is no likelihood that advanced Earth-like life could exist on its surface. For example, although receiving only half the average solar radiation of Earth (itself not an impossible limitation for photosynthesis) its eccentricity makes the solar radiation very changeable over the seasons. Moreover, as explained in Chapter 5, lacking a massive satellite like Earth's moon, the obliquity of Mars in its passage around the Sun, and hence seasonality, is unstable. This would be very challenging for the evolution of life.

The most life-limiting factors on Mars today are the almost complete lack of liquid water and an atmosphere. Even if a small amount of liquid water appeared on the surface it would quickly boil away under the very low atmospheric pressure. Gravity on Mars is only one-third that of Earth. This allowed the primeval water and atmosphere to slowly escape to space. In the distant past, at least 500 million years ago, Mars appears to have had a substantial atmosphere and the presence of surface water. This is supported by the appearance today of what seem to be dry river beds and extensive drainage systems.

Recent surveys by automatic probes have indicated the presence of frozen water at the Martian poles and possibly subsurface permafrost (the more visible polar ice caps are mainly frozen CO_2). Moreover, comparison of recent and earlier close-up pictures (separated by just a few years) indicates small surface flows of what is most likely water from subterranean geysers. If such outpourings do occur any liquid water would rapidly evaporate or freeze. These observations of possible liquid water have again raised the hope (fear?) that some form of microbial life may still exist. 'Fear' because we do not know if such microbes would be benign or dangerous to life on Earth, such as humans. If astronauts or soil samples return to Earth from Mars it is by no means certain that back contamination can be prevented. We are still very close to the 15th-century trauma of the meeting of Old and New World humans, which brought about a catastrophic exchange of pathogens. At that time a very large percentage of Native Americans died, as they were not immune to the smallpox, tuberculosis, measles, and other diseases introduced by the Europeans. There were also cross-infections in the reverse direction.

To date (early 2009) no unambiguous indication of residual life has been discovered, even at the level of single cells. But the search goes on, led by robotic probes prowling the Martian surface, directed from Earth with the aid of data received from them and a fleet of orbiting probes. A comprehensive review of Mars, as known up to 2001, can be found in Leovy (2001). Readers are referred to the NASA and ESA websites for updates which report the findings of Phoenix, the latest water- and organic molecule-seeking probe, which landed on Mars on 25 May 2008.

Venus

After Mars, Venus, is our nearest planetary neighbour (Fig. 3.5). Its size and gravity are close to those of Earth and its mass is 81% of that of Earth. There the similarities end. It is nearer to the Sun, at 0.72 AU (i.e. it is too close to be in the CHZ). It has a very dense atmosphere. The surface pressure is ~90 bar (that of Earth is, by definition, 1 bar). The composition of the Venusian atmosphere is 96% CO_2 and 3.5% N_2. The combination of a very thick atmosphere composed mainly of the greenhouse gas CO_2 and proximity to the Sun results in a surface temperature of 464°C. There is no liquid water on the surface, despite the high pressure. However, at an elevation of about 50 km the temperature of the atmosphere is within the 0–100°C liquid water range. This has raised speculation—some scientific and even more in the realm of science fiction—that life may exist, suspended at this elevation. To do so would require either floating microbes or 'floating bladders' adjusting their density to maintain their presence in this zone (Morowitz and Sagan 1967; Cockell 1999, 2005). This is not impossible; there are life forms in the oceans of Earth which do just that (although not in Earth's atmosphere). But note that the atmosphere of Venus is highly acidic, with drops of sulphuric acid at a concentration of 80–95%. There is no or very little water. This would be an osmotic and acidic challenge for even the most extreme of known Earth extremophiles.

Observations of the Venusian surface suggest that it is relatively young (of the order of 500 million years). At that time there seems to have been a catastrophic upwelling of magma, which not only destroyed all signs of older meteor impacts but would also have destroyed any form of life which may have emerged.

On Earth, tectonic movements contribute to the recycling of carbon and the control of CO_2 in the atmosphere. This moderates the greenhouse heating effect (Chapter 4). There are no such tectonics on Venus and the accumulation of CO_2 in its atmosphere led to runaway heating.

The physical and chemical characteristics of Venus are summarized by Beatty *et al.* (1999). Cockell (1999) reviewed the possibility of it harbouring life. The bottom line is that while life at the microbial level is not impossible it is most unlikely.

Summary

- To date (early 2009) the Solar System is the only star system where we *know* that life exists and there, only on Earth, within its narrow biosphere.
- Our star, the Sun, has certain characteristics which make it suitable for supporting the evolution of advanced life:
  - a longevity of ~10 Gy. Its present age is 4.8 Gy;
  - it formed from remnants of a supernova and as such contains metallic elements, in addition to hydrogen and helium;
  - it has a surface temperature of 5,770 K, which gives it a radiation that can be used by life forms on Earth.
- Earth's atmosphere attenuates the incident solar short-wave radiation by 50% and modifies its spectrum in a way which favours life, especially by absorbing most of the UV.
- Life, uses the Sun's radiation for heating, photosynthesis, and long-distance sensing, including vision.
- The circumstellar habitable zone (CHZ), a theoretical construct, is the band around a star in which planetary life is possible. The main consideration in defining the limits of the CHZ is the maintenance of planetary surface temperatures at a level which supports water in the liquid phase. CHZ models take into account the physical characteristics of the star and its planets and the presence and density of planetary atmospheres. Earth, at 1 AU, is almost ideally located for life in the CHZ, which is calculated to be between 0.8 and 1.6 AU.
- There are eight planets within the Solar System. Of these only Earth is well positioned in the CHZ and has other unique characteristics, favouring life.
- Within the inner Solar System there are three planets, other than Earth, that have solid surfaces (terrestrial type):
  - Mercury—close to the Sun and without a significant atmosphere; too hot and dry for life;
  - Venus—with a very thick atmosphere and very high surface temperature. At the elevation at which the temperature of its atmosphere falls to that which could support liquid water, it contains highly acidic droplets. Microbial life is possible at this level, but is most unlikely.
  - Mars—the planet nearest to and most similar to Earth. It differs from Earth in many ways that today prohibit advanced life: an unstable climate, very little atmosphere, and a surface too cold to support liquid water. However, 500 My BP it probably had a thicker atmosphere, water clouds, rain, and surface water. At that time it could have supported at least microbial life, remnants of which may survive today in protected niches. To date (early 2009) no unequivocal evidence of life has been found on Mars, ancient or extant.

4

The Solar System and life on Earth: II

Comets and asteroids

Comets and asteroids affect life mainly in a negative way—if and when they impact with Earth. Such catastrophic events are discussed later in relation to the evolution of life in an uncertain and constantly changing and challenging environment (Chapter 7).

Asteroids are essentially rocks, either hard or a loose conglomeration of dust and stony material. Some 75% are rich in carbon and are known as 'carbonaceous asteroids', 17% are silicaceous (S-type asteroids), and 8% are metallic (L-type asteroids).

Comets have been described as 'dirty snowballs' as they are composed mainly of water mixed with rocky material. Although comets are generally not good news for life on Earth (a belief of many ancient civilizations, extending to more recent times, at least as far as Shakespeare) one hypothesis suggested that comets were the source of Earth's water. However, today, only a small part of Earth's surface water is thought to have come from extraterrestrial sources (Chapter 5). Here we only briefly review the place of comets and asteroids in the Solar System as a prelude to understanding their effects on our biosphere.

In comparison to the inner reaches of the Solar System, which contains Earth and the other planets, the full extent of the Solar System in space is very great. Far outside the orbits of the planets lies the Oort cloud. This is a sphere of billions of comets at a distance from the Sun of some 100,000 AU or about 1 ly. One light-year is almost a third of the distance to Proxima Centauri, the star nearest to the Sun. The existence of the Oort cloud is still a hypothetical construct; it cannot at present be observed. When gravitational perturbations, possibly caused by a passing star, affect the Oort cloud, a comet may be jarred from its normal position. When this happens it may acquire a trajectory which brings it into the inner Solar System, where it orbits the Sun (Figs 3.5a and 4.1a). If it does it may be observed from Earth which, on this scale, is close to the Sun. An example is Comet Halley, which circles the Sun and can be seen from Earth every 76–79 years.[1] Comet Halley was visited by five probes, from NASA, the ESA, Russia and Japan, when it orbited the Sun between 1986 and 2001. It is predicted to return in 2061.

---

[1] The variability in orbital period is the result of the gravitational pull of other planets and the reaction to the gases it ejects.

Just outside the orbit of the planet Neptune, at a distance of 30–50 AU from the Sun, there is a band of some 70,000 objects called the Kuiper belt. Many such objects are large rocks, much larger than the comets of the Oort cloud. Although many of the Kuiper belt objects are of the order of 100 km in size, some are smaller comet-like objects. Gravitational perturbations by the larger planets may cause comets of the Kuiper belt to leave their normal orbits and circle the Sun. When they do, they too may be seen from Earth. With their shorter orbits they return every few years and are consequently called short-period comets (Fig. 4.1a).

Nearer to Earth, between Mars and Jupiter, in a band ranging between 1.7 and 4 AU from the Sun, lies the main asteroid belt (Fig. 4.1b). It contains rocky objects ranging in size from dust to ones so large as to be termed dwarf planets. More than a million may be more than 1 km across: among the larger are Ceres, at 1,003 km, and Pallas, at 608 km. The sky above Earth is continually assaulted by small rocks. Many are remnants of the formation of the Solar System, though some may come from the asteroid belts. They are attracted by Earth's gravity and enter the atmosphere, where friction causes them to heat and break up, leaving a characteristic 'fiery' trail, which is visible at night (as meteors or 'shooting stars'). The larger meteors may not completely disintegrate in the atmosphere but fall to Earth. They are then called 'meteorites'. Some of the more evident meteor showers are remnants of comets orbiting the Sun. They occur when Earth passes through a comet's tail.

Of special interest to biology are a few rare meteorites which contain magnesium minerals and also carbon compounds, including amino acids: the 'carbonaceous chondrites'. The most famous of these is the Orguiel meteorite which fell near Orguiel, France, in 1864. It split into 20 pieces with a total mass of about 12 kg. Among the many scientists who studied this chondrite was Louis Pasteur, who searched for but did not find any signs of extant microbial life. It is just possible that such meteorites originate from extraterrestrial sites containing at least some form of carbon-based life. However, although they contain hydrocarbons there is no other evidence that these compounds were formed in living cells. Their mineral composition suggests that they formed very early in the life of the Solar System.

Again, the main significance for life of comets and asteroids is that large ones impacting with Earth may have catastrophic consequences locally, and if large enough on a planet-wide scale (see Chapter 7).

Planetary satellites and life

There are numerous moons or satellites orbiting the eight solar planets. Jupiter alone has at least 63, Saturn 57, Uranus 22, Neptune 12, and more are still being discovered. One might expect them to be all rather similar to our own single, sterile, and meteor-impacted Moon; but this is not so. They are incredibly varied. The satellites orbiting the giant gas planets, like Jupiter, are themselves rocky in composition. Some are like the Moon—dry rocks, geologically quiescent, with no atmosphere. Others have

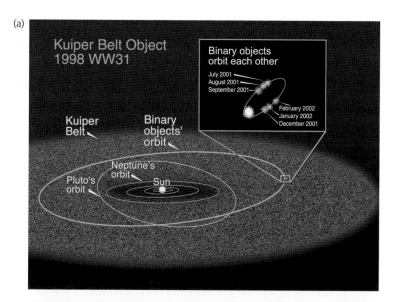

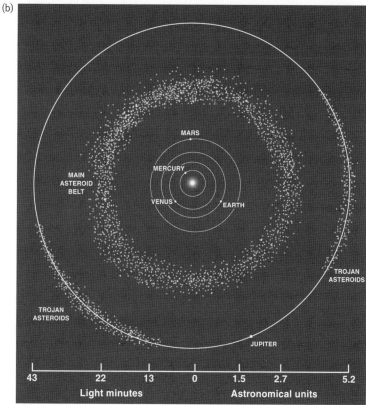

Fig. 4.1 (a) The Kuiper belt. (b) The inner asteroid belt. Note that these are not to scale. The Kuiper belt is at a distance of 30–50 AU from the Sun. The sphere of the Oort cloud with its many millions of comets (still a hypothetical construct) is thought to be at a distance of 100,000 AU. Courtesy of NASA.

atmospheres. At least one (Saturn's Titan) has both a dense atmosphere and liquid on its surface. Jupiter's satellite Io is about the size of Earth's moon but, unlike the 'dead' Moon, it is the most actively volcanic body in the Solar System. Europa, one of Jupiter's four Galilean moons (named after their discoverer) is completely covered in water ice.

Some of the planetary satellites are causing considerable excitement amongst astrobiologists. They are all outside the CHZ and too far from the Sun for photosynthesis, but they have other sources of energy, as described below, and some may have warm niches, warm enough for liquid water. Some are extremely cold but have molecules, such as methane, which are liquid at their surface temperatures. Study of these bodies may shed light on conditions which existed on early Earth and engendered life. If we entertain the panspermia hypothesis, seeds of life could have wandered from them to Earth (or the reverse). Some may be found to have life forms adapted to their special environments (sometimes called 'non-terran life'; Anonymous 2007). Moreover, for astrobiologists interested in the search for extraterrestrial life in the universe at large, they may well serve as model systems and expand the theoretical boundaries of the GHZ and CHZ (see Chapter 2).

Many of these satellites are believed to contain aquifers of liquid water, perhaps as large as oceans. Within the Solar System today, although ice is quite common, only planet Earth has a persistent presence of liquid water on its surface.

Three examples of these fascinating satellites are presented. Research on these bodies is so active that no book or printed review can be up to date. For the latest advances, outside of printed reports in the science literature, readers are referred to the NASA and ESA websites and Wikipedia (see References and Resources). However, bear in mind that this latter source receives only limited editorial review and mistakes do sometimes appear (which is also true of peer-reviewed science papers!). For 'weird life', meaning possibilities for life which is unlike that found on Earth, see a recent report by a special commission of the US National Research Council (Anonymous 2007).

Europa—an arctic look-alike (?)

Europa is the fourth largest of Jupiter's many satellites. It is the smallest of the four Jovian moons discovered by Galileo. Much of what we now know of Europa comes from telescopes on Earth, the Hubble space telescopes and NASA's Galileo probe. The latter, which orbited Jupiter flying close to Europa, in a number of fly-bys between 1995 and 2003 (Fig. 4.2).

The radius of Europa is about a quarter that of Earth. It is thought to have a metallic iron core covered by a silicate rock layer. Its gravity is only one-seventh that of Earth. It has very little atmosphere (atmospheric pressure at the surface is ~ 1 μPa compared with Earth's 10^5 Pa) which is almost certainly oxygen from a non-biogenic source. What makes this moon so interesting is that its surface is entirely covered with water

Fig. 4.2 Europa, a Jovian moon with an ice-covered ocean. Note the line markings which are thought to be deep cracks in the water ice covering the satellite. Courtesy of NASA.

ice. Its surface temperature varies from 50 to 125 K and the ice apparently overlays an ocean of liquid water thought to be 100 km deep. This may be compared with the Arctic Ocean, below the ice fields of Earth's northern latitudes, which supports a relatively rich ecosystem.

The source of the heat required to maintain the temperature of this putative ocean so far from the Sun (at 270 K at least[2]), is thought to be the tidal, gravitational pull of the nearby massive planet Jupiter. Added to this is the effect of irregular motions of Europa resulting from the gravitational perturbations of other Jovian moons.

Hand and Chyba (2007) analysed the information gathered by the Galileo probe and all other known data on Europa. While recognizing the present paucity of data they conclude that the best estimate for the thickness of the ice is less than 4 km. The sub-surface ocean would be very salty. However, there are microbes on Earth which live at such temperatures and salinities. For a review of the 'habitability' of Europa see Sotin and Prieur (2005).

Preliminary plans exist to land a nuclear-powered probe on Europa, which would penetrate the ice and search for life; plans, but not budgets. It will probably be decades before the results of *in situ* exploration can be expected.

Titan—a moon with a dense atmosphere and lakes

Titan is the largest of the moons of Saturn, with a radius of about 2,575 km, 40% that of Earth. It was first observed by the Dutch astronomer Christiaan Huygens in 1655. Following more recent telescope studies from Earth and space (by the Hubble tele-scope), our knowledge of Titan has been augmented tremendously by data returned from NASA's Cassini probe, which arrived near Titan in 2004. As of 2008, this probe is still making data fly-bys over the satellite, after dropping its Huygens lander probe (built and operated by the ESA) in January 2005.

The gravity of Titan is only one-seventh that of Earth, but even so it has a dense atmosphere with a surface pressure of 146.7 kPa, 1.5 times that of Earth, which is prob-ably being constantly replenished by degassing. In this latter respect (and in many other characteristics) it is unique among the moons of the Solar System. The compos-ition of the atmosphere is mainly nitrogen (98.4%) and methane 1.6%. It has clouds, rainstorms, and near its north pole even lakes. Size and density studies suggest that it is formed of about 50% rocky material and 50% water. Part of this water may be in the form of an underground liquid layer, or ocean.

Given just the above description, which fits everything known of Titan up to early 2009, we could almost be excused if we expected to see little green men flying happily through the atmosphere of their low-gravity home. But a closer look shows a world and environment entirely different from our own. No terran life could survive the conditions

---

[2] This minimum is a little below 273 K (0°C) as the sub-ice ocean is thought to have a considerable salt content.

on Titan, apart perhaps from microbes in very small and specialized niches. Titan's atmosphere is very reflective of short-wave solar radiation (i.e. it has a high albedo). This tends to counteract the greenhouse effect of its atmosphere. The surface temperature on Titan is only 90 K above absolute zero (0 K), far too low for liquid water. Titan's clouds, rain, ponds, and lakes (the latter at its north pole, with some estimated to be of the size of the North American Great Lakes or larger) are formed mainly of hydrocarbons, with a preponderance of methane and ethane. These hydrocarbons, which are gases at Earth temperature, are liquids on Titan.

Any life which evolved on Titan would have a chemistry quite different from that with which we are familiar. For example, heat-driven metabolism would proceed at a fraction of its speed on Earth, with life forms having a life span of perhaps hundreds of Earth years (a great advantage for space flight—see Chapter 10). Our imaginations can run wild; but here we delve into science fiction! More serious consideration and a workable hypothesis for life on Titan have been offered by Schulze-Makuch and Grinspoon (2005) and by Raulin (2005).

Enceladus—a tiny but intriguing world

Enceladus (Fig. 4.3) is a very small satellite of Saturn. It was first discovered by the British astronomer William Herschel in 1789. Its radius is only 250 km, about one-seventh that of Earth's moon, with a proportionally low gravity (0.012g). Modern studies of Enceladus began with the Voyager probe of the 1980s, which returned pictures showing a very varied terrain, some tectonically formed. Astrobiologists were excited by pictures and data returned in 2005 by the same Cassini probe described above, which arrived in the Saturn system in 2004. In 2005 it made three fly-bys of Enceladus, the nearest at a distance of 175 km. It continues to periodically approach Enceladus, and six fly-bys were made in 2008, three of which were at a distance of ~23 km.

The data already gained have been so fascinating as to warrant an issue of *Science* almost entirely dedicated to Enceladus (introduced and summarized by Baker, 2006). Its surface is icy like Europa but it also has exposed rocky mountain folds. There are signs of active tectonic motion and relatively new ice flows striped with what appears

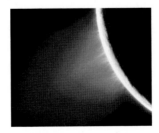

Fig. 4.3 The south pole of Enceladus, a tiny rock with water fountains. Courtesy of NASA.

to be dark green organic(?) material. The area of the south pole is of greatest interest. In this region Cassini observed an out-gassing of a cryo-volcano. The quantity of water and ice ejected is so great as to replenish Saturn's entire E-ring, within which Enceladus orbits the planet. There is no doubt that the cryo-volcano is being supplied by an internal, relatively hot pool of subsurface water. The nature of this aquifer and what maintains it at a temperature above freezing (273 K) is unknown. Gravitationally induced tidal forces are no doubt involved, as on Europa.

Again, these studies of Enceladus demonstrate that there are at least a few extraterrestrial sites in the Solar System where liquid water can or does exist. They may have survived for long periods of time, possibly providing an ecological niche for the emergence and evolution of non-photosynthetic, microbial life.

Earth—a unique planet of the Solar System

It is somewhat platitudinous and certainly a tortuosity to state that Earth has the special conditions enabling life. Of course it does; otherwise we would not be here to discuss them. Having admitted to this logical weakness, we can still address the special properties of our planet. They, affected in turn by the Solar System and the extra-Solar System factors discussed above, have enabled the appearance and evolution of life in the biosphere.

Earth is the third planet from the Sun, positioned almost ideally within the theoretical CHZ (see Chapter 3). Some of its main physicochemical characteristics were discussed earlier in relation to and in comparison with Mars (see Table 3.1). Its main life-enabling characteristics are a temperature which allows water to exist in a liquid state on much of its surface, and an atmosphere. The radiation-absorbing properties of the atmosphere support the surface temperature (Fig. 3.4) and its gaseous pressure suppresses the rapid evaporation of liquid water. Its pristine composition allowed for the emergence of anaerobic, unicellular life. Its present composition supports advanced, multicellular life.

The surface environment of Earth shows long-term *relative* stability. This is of great significance for the evolution of advanced life forms. Conditions on Earth have allowed life to emerge, survive, and to continue to evolve over a period of 3.7 Gy. This last statement must be qualified by the many environmental changes, some calamitous (but never *completely* destructive of life), that life has experienced on Earth (see Chapter 7). The special and unique life-supporting geological properties of Earth are discussed below.

Size, mass, and gravity

As forces go in nature, gravity is very weak. It is 26 to 40 orders of magnitude less than the strong or weak nuclear forces or the EM force, which together connect atoms and particles in nature. Even so it is of the utmost importance at the cosmic scale, affecting the scatter of stars and galaxies and even the path of light in space. At the scale of the Solar System it affects the distribution and orbits of the planets, comets, and other bodies, and their interactions with each other.

On Earth the acceleration due to gravity, g, has a value of 9.81 m s^{-2}. This represents the acceleration of a body freely falling in a vacuum, and is a measure of the 'strength' of the gravitational force. Gravity is determined by the mass and size of a body An important effect of the strength of the Earth's gravity is that it enables Earth to hold on to its atmosphere, only a little of which escapes to space, being replaced by out-gassing of volcanoes. At the same time it is not so strong as to attract a denser atmosphere which, with respect to the distance of the Earth from the Sun, would produce a surface temperature unacceptable for life.

Rayner (2003) has discussed the suggestion that the value of g may have changed considerably over Phanerozoic time as a result of a putative change in Earth's volume. However, we go along with his tentative conclusion that firm evidence is lacking. If, however, the value of g has changed, it would have had enormous repercussions on the environment and the evolution of advanced life.

Earth rotates, and the resulting centrifugal force at the equator makes it bulge, producing a slightly oblate sphere. Earth's equatorial radius is 6,378.14 km and its polar radius is 6,356.75 km. This difference may seem negligible, but the slightly lower gravity at the equator (as a result of the greater distance to the Earth's centre of mass) together with the centrifugal force resulting from Earth's rotation, affects the weather. This is because the atmosphere is less dense, and cloud formations (especially cumulonimbus storm clouds) reach great altitude and are far more energetic than at higher latitudes. In addition there is the extra incident solar radiation as a result of the higher solar angle above the horizon. These, together with other geographical factors, give rise to heavier precipitation events in low-latitude regions. Life on Earth has evolved while accommodating to and utilizing gravity. Here are a few examples:

- Some algae, which live in water, have flotation mechanisms which adjust their buoyancy so that they float at the optimum depth. 'Optimum' refers here to light intensity, temperature, nutrient supply, etc. This entire mechanism is of course gravity dependent.
- Germinating seedlings of higher land plants utilize gravity in two opposing ways: their emerging roots grow in the direction of the gravitational force ('positive gravitropism'), ensuring their penetration into the soil and their shoots grow in the direction opposite to the gravitational force ('negative gravitropism'), i.e. upwards in the direction of light, required for photosynthesis.
- The entire geometry of advanced land plants on Earth is adapted to gravity. As plants grow higher they must expend more resources on physical support (e.g. lignin in tree trunks) to overcome their weight. This limits the height of land plants to about 150 m. The tallest extant tree known today is a redwood in Mendocino, California, USA, which stands at 113 m. Remains of Australian eucalyptus trees have been found

with a calculated height of about 150 m. The geometry of leaves is also dependent on structural mechanics to overcome gravity and wind forces (Niklas 1998).

- The structure of animals, particularly their bone strength, is a function of gravity. This means a considerable allocation of resources to bones. Our bodies tend to adjust to changing gravity. During microgravity flight in spacecraft the bodies of astronauts conserve resources by reducing bone mass. As we age our body structures tend to sag. Repairing ravages of the gravitational force is one of the main *raisons d'être* of the multi-billion dollar plastic surgery industry!

- Bird flight evolved to overcome gravity and allow movement through the air of heavier-than-air bodies. If only there were significant atmospheres, we massive humans would be able to fly on Mars (where gravity is only 0.38g) or on the Moon (0.165g) with only angel-sized wings! The story of human heavier-than-air flight, since that of the Wright brothers in the USA in 1903 (and probably Dumont in France in the same year), is an epic in the human struggle to overcome gravity. The effects of gravity on life have been reviewed by Morey-Holton (2003) and by Rayner (2003). Life without gravity or in microgravity environments has been studied to some extent in space stations and in the Space Shuttle. Gravity appears to be essential for many plant and animal processes. This is not surprising as life evolved under conditions of 1g. Astronauts returning to Earth after prolonged space flights ($\sim$1 year) under conditions of microgravity have had severe health problems, despite numerous on-board treatments, such as nutrition and exercise, to counter such effects (Wassersug 1999).

The structure of Earth

The analysis of seismic waves passing through the Earth during earthquakes has revealed a reliable picture of the internal structure of the Earth (Fig. 4.4).

The inner core of the Earth is very dense ($\sim$13,000 kg m^{-3}). It is thought to be composed mainly of iron and nickel. Although very hot, at about 7000 K, the high pressure, of the order of 360 GPa, keeps it solid. The source of the heat is the decay of radioactive isotopes of potassium, uranium, and thorium. Their half-lives are in excess of a billion years. This heat diffuses outwards. It makes a small contribution to the temperature balance of the Earth's surface. It also provides the heat and, via certain reducing molecules, the energy used by the very significant mass of microbes which inhabit the outermost 3 km thick layer of the planet (see Chapter 8).

The outer core is liquid, and circulates. This is thought to give rise to the magnetosphere which surrounds the Earth (and to different degrees some, but not all, bodies of the Solar System which have similar iron and liquid cores). Its significance to life is less certain and is discussed further in Chapter 5.

Above the outer core is the mantle, which is composed of hot, almost solid rock. Its outer layer, the asthenosphere, flows like a liquid, but extremely slowly. The upper mantle and outer crust (together making the lithosphere) are made of plates of solid

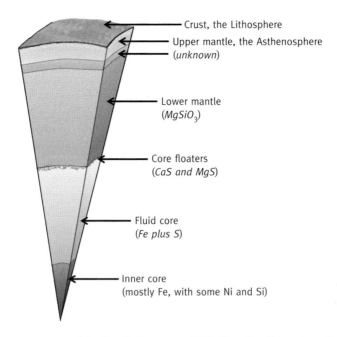

Crust, the Lithosphere

Upper mantle, the Asthenosphere

(*unknown*)

Lower mantle
(*MgSiO₃*)

Core floaters
(*CaS and MgS*)

Fluid core
(*Fe plus S*)

Inner core
(mostly Fe, with some Ni and Si)

Fig. 4.4 The inner structure of the Earth. Courtesy of J.M. Herndon, Transdyne Corp.

rock. The movements of the lithosphere over the mantle (plate tectonics, see below) are the source of numerous phenomena that strongly affect life.

The paradigm of plate tectonics

The apparently complementary shapes of the continents on the two sides of the Atlantic had long intrigued cartographers. Savants such as Francis Bacon and Benjamin Franklin had hypothesized that the land masses had once been joined. This idea, called continental drift, was also supported by the American, Frank Taylor, at the beginning of the 20th century. The first to present evidence that the continents had indeed drifted apart was Alfred Wegener, who published his findings in 1912 and continued his work to 1930. By placing together geological maps of the two sides of the Atlantic, Wegener showed numerous palaeotopographical coincidences and continuations. However, in the absence of a mechanism for such an ostensibly outrageous proposal, his ideas were summarily dismissed and forgotten for half a century.

A mechanism for tectonic movement had been suggested by Arthur Holmes, in 1929, but was generally rejected. Only in the 1950s, with the discovery of magnetic anomalies around the mid-Atlantic ridge, was Holmes' hypothesis proven and Wegener's proposals vindicated. These anomalies appear in the form of symmetrical north to south stripes on either side of the mid-Atlantic ridge. They are interpreted as showing sea-floor spreading, resulting from upwelling magma which, as it cools, becomes magnetized in the direction of the prevailing magnetic field. Sea-floor spreading was the missing mechanism for continental drift.

From the 1960s this hypothesis developed into the overall paradigm of plate tectonics, which includes continental drift, sea floor spreading, and subduction zones. The modern theory is based on an understanding of the mechanics of the system. The lithosphere is now known to be divided into some seven major and many minor plates. They are approximately 100 km thick in mountainous regions, considerably less under the oceans. These plates ride on the asthenosphere, the more viscous part of the outer mantle. Plate motion is slow, from a few millimetres to some 15 cm a year. Where plates meet, three different types of boundaries are recognized: (1) divergent, where the plates are slowly sliding apart; (2) convergent, where plates collide; and (3) transform, where the plates are slowly sliding past a common boundary, in opposite directions.

The undersea rocks are relatively young (less than 180 My) while on dry land some are 4 G years old; almost as old as Earth itself (4.55 Gy). This is because the undersea rocks are denser than the continental rocks. Hence, when two plates converge, one which has oceanic rocks will sink below one made of continental rocks. The subducted rocks melt and are mixed within the magma. This also happens when two oceanic plates converge. When two continental plates converge, neither sinks and the plates buckle up forming massive mountain ridges. The Himalayas formed as a result of such a collision between the Indian and the Eurasian plates. As noted, the undersea plates are much denser than the new plate material formed by the magma exuding from the somewhat elevated ridges. Consequently, the new plate material tends to slide down the ridge, producing an expanding plate movement, as it spreads away from the ridge. As the plates age they become denser; causing them to sink at subduction zones. This is one of the major driving forces of plate movement.

Starting from about 1960, this theory has been supported by numerous findings and it has produced major advances in our understanding of geological processes. Plate tectonics explains many of the formerly enigmatic characteristics of Earth. These include mountain building, the presence of sea floor rock assemblages which include fossils of sea life at high elevations, earthquakes and tsunamis, and volcanoes. The consequences of plate movements are of great importance to our understanding of some of the factors affecting and perhaps effecting the evolution of life (Forster 2003).

Since the relatively recent (in evolutionary terms) Triassic period, some 200 million years ago, the land mass of Earth has broken up from a single continent called Pangaea to what we know today (Fig. 4.5). Note that this break-up of the continents is a relatively recent event in the 4.55 billion year history of Earth. This was not the first time that the land masses had been first consolidated and then separated. The continents had previously been joined some 880 million years ago.

Tectonic movements explain earthquakes, which are 'shudders' of the plates, especially where they suddenly move in opposite directions. This occurs along plate transform boundaries, after a period during which friction kept them locked and tensions built up. Two well-known examples of such earthquake zones along plate boundaries are the San

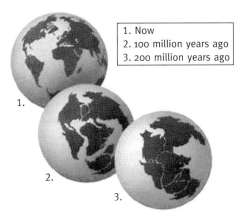

1. Now
2. 100 million years ago
3. 200 million years ago

1.

2.

3.

Fig. 4.5 Continental drift since the Permian period.

Andreas Fault in California and the much larger Great Rift Valley, which extends from Africa to eastern Europe. Sub-oceanic earthquakes are often associated with tsunamis.

Where plates collide at convergent boundaries there may be subduction, with one plate sliding below the other. As explained below, this may have been of great importance in the past, when subduction activity contributed to the reduction of atmospheric carbon dioxide.

At borders between plates there is often a release of magma, water, and gases from deep in the Earth. Such occurrences may be as minor as hot springs, as fascinating as deep sea vents (Chapter 8), or as dramatic as volcanic eruptions. The latter have sometimes been on a colossal scale, with world-wide catastrophic consequences to the biosphere (Chapter 7). The eruption of an Indonesian volcano about 70,000 y BP is though to have brought the entire world population of modern humans close to extinction (Chapter 9). The whole of the Pacific Ocean is bounded by numerous volcanoes, sometimes called the 'ring of fire'. Many volcanoes are active today. In human historical time their eruptions have produced local catastrophes, for example Vesuvius, in Italy, which destroyed Pompeii in 79 CE. The 1883 eruption of Krakatoa, in Indonesia, killed 36,000 people and affected the climate on a world-wide scale. The dust it spewed into the atmosphere reduced solar radiation (and hence plant growth) world-wide, for 2 years. The 1991 eruption of Pinatubo in the Philippines was almost as large, and was one of the largest volcanic eruptions of the 20th century.

Movement of tectonic plates may produce changes in local environmental conditions. For example, sea currents and climate of adjacent dry land were very different in the time of Pangaea. Today, some northerly land areas, close to the sea, such as the coast of Norway, have their climate moderated by warm sea currents flowing from regions close to the equator. Climate change resulting from the realignment of the continents may induce migration, adaptation, or death of living species.

There are a number of effects of plate tectonics which are so significant that some authorities believe that they were essential to the evolution of life. Among these are its consequences for speciation, defined as separation into non-interbreeding types with distinct characteristics. Under Pangean conditions life forms would tend to mix and interbreed freely and fewer separate species would form. Diversification of species and enriched gene pools are considered to be important and perhaps essential, for meeting the challenges of Earth's constantly changing environment.

Another argument in favour of the critical importance of plate tectonics for life is by way of a 'thought experiment'. It has been calculated that if plate movements were to stop, the land would slowly erode into the oceans. The level of the seas would rise and cover most of today's dry land, albeit with a relatively shallow ocean. Only a few, small and isolated islands would remain as dry land. Would intelligent life have arisen in the seas? We can only speculate, with dolphins, which originated on dry land, being our best example. However, this calculation needs further corroboration.

Perhaps the strongest argument in favour of the essential importance of plate tectonics to advanced life is in relation to the stabilization of Earth's temperature. On the early Earth, 4.55 billion years ago, the intensity of the Sun's radiation was about 25% weaker than today.[3] The problem is to understand how the Earth maintained its more or less constant surface temperature over billions of years, since the first appearance of life around 3.7 billion years ago (cf. the faint early Sun paradox mentioned earlier). Even a 0.25% increase in solar radiation, as sometimes occurs during periods of sunspots or in response to a decrease in Earth's albedo, may cause a significant ($\sim$1°C) increase in world temperature. That there was no catastrophic heating (despite the evidence for many periods of alternating warm periods and ice ages; see Chapter 7) is shown by the record of life itself.

Life has experienced many challenging environmental catastrophes during its 3.7 billion year history, but evolution never stopped entirely and started again. This means that there was always liquid water somewhere on Earth. Consequently temperatures must have remained within the range 0–100°C (for liquid water) and most probably around 20 ± 15°C (the range to which most life forms adapted). As described below, the atmosphere of the early Earth was rich in CO_2, comprising around 3% of the atmosphere (before the Industrial Revolution it comprised 0.025%). This is thought to have produced a strong greenhouse effect (see Chapters 3 and 10) which allowed the surface temperature to reach levels suitable for life despite the relatively low solar radiation. With the evolution of photosynthetic organisms (Chapter 6) most of this CO_2 was removed from Earth's atmosphere and fixed in plant tissue.

Much of this organic material became buried in the soil, without being fully decomposed by microbes, a process which returns CO_2 to the atmosphere. This was (and is

---

[3] The circumference of the early Sun was about 6% smaller and its surface temperature some 300 K cooler than today.

today) the case in colder regions of Earth, where peat often accumulates. In the warm tropics there is a much more complete breakdown of plant remains. Where the organic material is not fully decomposed it may become buried under conditions of heat and pressure, ending up as coal. In some regions subduction processes acting on buried organic material produce both coal deposits and the carbon-rich carbonatite rocks. The subducted plant material may be converted into coal. Carbon dioxide is thus removed from the atmosphere until returned by the action of humans mining and burning fossil fuels.

Some of the CO_2 is removed from the atmosphere by diatoms. Diatoms are eukaryotic single-celled algae which have unique cell walls formed of hydrated silicon. Dead diatoms are relatively dense and they tend to sink in the seas, forming silica, carbon, and calcium-rich sediment. Subduction tends to bury these sediments together with plant remnants. The latter, under the effect of heat and pressure are sequestered deep in the Earth, forming coal and oil. (This is only one possible mechanism of oil formation. The origin of oil is still not well understood.) In this way CO_2 is removed from the atmosphere, reducing the greenhouse effect and thus mitigating the effect of the increased solar radiation.

At the same time as photosynthesis was removing CO_2 from the atmosphere volcanoes were adding CO_2. As described in Chapter 6, only the *combined* effects of oxygenic photosynthesis and carbon-sequestering tectonics produced the low-CO_2, high-oxygen atmosphere which enabled the evolution of advanced life.

Apart from photosynthesis, a major factor in the above process which removes CO_2 from the atmosphere is the weathering of silicates such as feldspar (a mixture of mainly $CaSiO_3$ with aluminium, potassium, and other silicates). In this reaction water-insoluble calcium carbonate and silicon oxide are formed and sink to the ocean floor. In a seminal paper, Walker (1994) described a feedback system which could control the temperature of the Earth by way of this reaction and the action of tectonics (Fig. 4.6). As shown in this figure, according to the above model carbonate deposits are formed on the ocean floor from the combination of atmospheric CO_2 and calcium silicates (produced by weathering). The carbonates are then sequestered deep in the Earth by subduction resulting from the movement of oceanic plates. There they slowly decompose (reverse weathering) releasing CO_2, which returns to the atmosphere, and calcium silicate, which becomes available for re-fixing CO_2.

As the Earth's temperature rises, the rate of weathering increases. This lowers the concentration of CO_2 in the atmosphere and tends to reduce the greenhouse effect and global heating. Then, as the Earth cools, the rate of weathering (carbonation of the silicates) decreases. Consequently the CO_2 content of the atmosphere increases and Earth temperature tends to rise. There are, however, newer studies which suggest that the rise in temperature and rainfall is less important than the appearance of fresh surfaces as a result of active plate tectonics (Riebe *et al.* 2001).

The above system is elegant in its explanation of the maintenance of Earth's apparently almost constant temperatures but, to this author at least, perhaps a little too

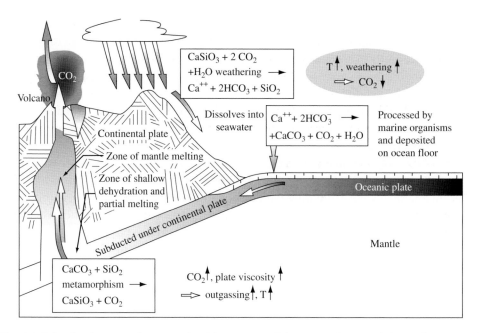

Fig. 4.6 A feedback system between the CO_2 content of the atmosphere and the carbonation of silicates, stabilizing Earth's surface temperature. From J.I. Lunine (1998). Courtesy of Cambridge University Press.

facile. The approximately 20% rise in the intensity of solar radiation over the last 3.7 billion years (since the appearance of life) was monotonic, a result of physical processes within the Sun. However, changes in the atmospheric CO_2 sources and sinks, such as CO_2 outgassing in volcanic eruptions and tectonic plate movements over the same time period, fit no obvious pattern. The oxygenic photosynthetic reduction of atmospheric CO_2 commenced only about 1 billion years after the appearance of life, at the end of the Archaean era. Atmospheric CO_2 was high in the Cambrian period and again in the Triassic, but low in the Permian and again in the Cenozoic. Throughout this time there have been periods of ice ages and interim warmer periods with no clear, recurring time pattern. Tectonics and the temperature feedback control system outlined above (Fig. 4.6) may contribute to stability, but the control cannot be close and immediate. However, for the reasons outlined above it is undeniable that Earth's temperature has remained remarkably constant throughout the 0.5 billion years of the Phanerozoic era. As described in the Walker *et al.* (1981) model and by Ditlevsen (2005), tectonic activity was and is probably a major contributing factor.

The argument which supports the importance of tectonic activity for life on Earth is relevant to the search for extraterrestrial life. Many astrobiologists believe that if there is no tectonic activity on a given cosmic body, oceans and dry lands will not come into being and surface temperatures will not be constant over the long time period required for the evolution of advanced life. Tectonic activity is currently absent on the other planets and most planetary satellites of the Solar System. There may have been such activity on Mars during its distant past and today it appears to be present on some of the planetary satellites.

Why tectonic activity is uncommon in the Solar System is not well understood. On Earth it may be associated with the huge quantities of liquid water (Ward and Brownlee 2000). As of 2008 there is no way of knowing if tectonic activity is present on any of the 250 plus extra-Solar System planets discovered in the last two decades. Then again, we are not absolutely sure that plate tectonics are essential for maintaining conditions favourable for life.

Cosmic radiation and the solar wind

The term 'cosmic radiation' is something of a misnomer. First it is not really 'radiation' in the sense of EM radiation, such as the short-wave radiation of the Sun or the long-wave radiation emanating from Earth (discussed above). Rather it is a stream of particles, some of which reach Earth's upper atmosphere with a very small, but biologically significant fraction, penetrating to the biosphere. Cosmic radiation is composed mainly of protons (~90%) and helium nuclei (~8%), also known as alpha particles. The remainder is electrons and positrons. The energy levels of cosmic radiation are extremely variable. One important source of such radiation is not cosmic at all, but comes from our Sun, in what is called the solar wind. The latter particles are usually of much lower energy (10^6–10^8 eV) than those originating from outside the solar system (10^8–10^{19} eV). Most of the latter appear to come from explosions of supernovae, with the most energetic cosmic radiation coming from unknown sources outside the Milky Way galaxy.

The solar wind interacts with and attenuates the radiation of cosmic origin, first at the very edge of the Solar System, the heliopause (see Fig. 3.1), and then again within the Solar System. The degree of attenuation depends on the flux of the solar wind, which varies with the 11-year solar cycle. When cosmic radiation reaches Earth it is deflected by the magnetosphere (see Chapter 5 and Fig. 5.2). Most of the remainder is absorbed by the atmosphere, Earth's most important shield. Without this protection cosmic radiation, which is energetic enough to cause severe ionization, would soon sterilize Earth.

Only a small fraction of the cosmic radiation penetrates the magnetosphere and Earth's atmosphere. However, the cosmic radiation that reaches the biosphere plays an extremely important role in life. As described in Chapters 2 and 8, evolution is driven to a large extent by the appearance of mutations, errors in the DNA/RNA information system. These errors tend to be perpetuated from one generation to the next. Ionizing radiation, such as high-energy cosmic radiation, is an important driver (albeit not the only one) in the production of such errors. Whereas the vast majority of such errors are of no value, or are even deleterious or lethal, and tend to be eliminated, a very small number produce a survival advantage and endure. These advantageous mutations are one of the important factors in adaptive evolution.

Microbial systems exposed to greater fluxes of cosmic radiation in spacecraft experiments have shown a variable response. Some have been damaged at the molecular information (DNA/RNA) level, while some have been remarkably resistant. The latter are of interest in relation to the possibility of 'seeds' of life surviving inter planetary and perhaps extra-Solar System space journeys—the panspermia hypothesis mentioned before (Horneck 2003).

Perhaps the most important indirect effect of cosmic radiation on life is by way of the weather. Shaviv (2003a) suggests that this effect may be the solution to the faint early Sun paradox. The young Sun had a stronger solar wind. According to Shaviv the solar wind reduces the flux intensity of the high-energy galactic cosmic radiation which penetrates the atmosphere. Cosmic radiation which does enter the lower atmosphere induces nucleation of water vapour and cloud formation. With the gradual reduction in the solar wind there would be more cosmic radiation reaching Earth, more cloud formation, and hence a greater atmospheric albedo. This would reduce the short-wave solar radiation reaching the surface of Earth and in this way would counter the effect of the increased short-wave solar radiation at the top of the atmosphere (the solar constant). This cooling effect would be in addition to the reduced greenhouse heating resulting from the reduction in atmospheric CO_2 content described above.

The Solar System moves within its 'home' the spiral Orion arm of the Milky Way. As it passes through the denser central part of the arm it experiences gravitational perturbations which affect the cosmic radiation environment. This periodic passage, together with variations in the rate of star formation, is thought to be responsible for some of the vagaries of Earth's weather, including ice ages, on a scale of decades, centuries, and millennia (Shaviv 2003b). This is further discussed in Chapter 9, in relation to human versus natural causes of climatic change.

There are at least two important, indirect biological effects of the solar wind. One is on communications. During solar flares—periods of intense solar wind—which may last from minutes to days, there is often severe interference with the functioning of electronic communication satellites. These orbit Earth well outside the protection of the atmosphere.

Another effect of cosmic radiation of considerable practical importance is in carbon dating. This is a very widely used and important dating method in biology, archaeology, earth sciences, and numerous other disciplines. It is based on the relative abundance of ^{14}C, the radioactive isotope of carbon, versus the common and stable ^{12}C. The radioactive ^{14}C isotope is formed in the atmosphere by the action of cosmic radiation on ^{14}N and is assimilated along with ^{12}C by living organisms during their lifetime. It has a half-life of 5,730 years, and reverts to ^{14}N by beta decay. Analysis of the relative abundance of the carbon isotopes in a sample indicates the time since the carbon was free in the atmosphere. This method is good for dating as far back as 60,000 years ago (at the very most), but fails for older dating owing to our lack of knowledge of the flux stability of cosmic radiation over long periods of historic time. Moreover, after eight half-lives (i.e. 46,000 years) only 1/256 of the original ^{14}C atoms remain. This makes detection very difficult against the normal background radiation.

Summary

- Asteroids and comets are remnants from the time of formation of the Solar System. Asteroids are solid (as opposed to gaseous) bodies. Comets contain solid material and a high proportion of frozen water. Millions of comets are found some 100,000 AU from the Sun in the Oort cloud. A few are in the Kuiper belt, at a distance of 30–50 AU. Occasionally comets wander into the central Solar System. Asteroids are found in the Kuiper belt, and in the main asteroid belt, at a distance of 1.7–4.0 AU from the Sun. They range in size from loose bundles of dust to rocks hundreds of kilometres wide. Comets and asteroids which leave their normal orbits and impact Earth have been, and may be, very destructive to life.
- Some of the planetary satellites (moons) may have conditions that could support either Earth-similar extremophiles or highly exotic life forms. Among these are Europa, a Jovian moon covered with water ice, under which lies a liquid ocean, perhaps 100 km deep, and Titan, a moon of Saturn, which has an atmosphere, clouds, rain, and lakes, formed mainly of liquid methane and ethane. If life has emerged on Titan it would not be based on water. Enceladus is a very small moon of Saturn. It contains liquid water, which is ejected into space in the form of ice and water vapour.
- The physical size, mass, gravity, and distance from the Sun and other chemical factors, are the characteristics which imbue planet Earth with the conditions which enabled and continue to support life as we know it. It is thus *the* model in the search for possible life-supporting extraterrestrial cosmic sites. Its central core is solid, mainly nickel and iron. The outer core is liquid. Its circulation is probably the source of Earth's magnetic field. Further from the core is the mantle which is liquid but very viscous. Sometimes it behaves like a solid, allowing passage of detectable seismic waves. On it floats the lithosphere which includes the outermost part of the mantle and the crust, which is divided into several rigid plates. The movement of these plates explains most of the geological characteristics of Earth, such as mountain formation, volcanoes, earthquakes, and tsunamis. Some consequences appear to be very important, if not vital to life, such as the relative areas of dry land, islands and oceans (important to speciation and affecting world climate); build-up of mountains; burial of plant debris (which changed the atmosphere's composition and produced fossil fuel); and, of greatest importance, the contribution to Earth's relatively stable surface temperature over aeons of time.
- Cosmic radiation is particulate radiation (mostly protons) which arrives at Earth from outside the Solar System and from the Sun (the Solar Wind). Cosmic radiation has many effects on life. Weather is affected by the induction of cloud formation, which also increases Earth's albedo tending to decrease its temperature. This may have been an important factor in the stabilization of Earth's

temperature as the Sun's radiation gradually increased over the lifetime of the Earth (the faint early Sun paradox). The small amount of cosmic radiation which penetrates the atmosphere and reaches the biosphere is highly energetic and ionizing. It induces mutations in living cells. The tiny fraction of these mutations which are beneficial are the basis of evolution.

- Cosmic radiation has some important practical effects on human life. For example, it is the generator of the ^{14}C isotope, used in numerous dating applications. Sudden increases in cosmic radiation flux (usually from solar flares) interfere with electronic communications.

5

Early and present Earth and its circumnavigation of the Sun

Early Earth

The Solar System came into being from the remnants of a supernova event around 4.55 billion years ago. As discussed in Chapter 2, this supernova origin was important for life as it ensured the presence of the essential heavier than helium elements, such as carbon, iron, and phosphorus. The first period of Earth's formation, from 4.55 to ~4.0 Gy BP is called, appropriately enough, the Hadean aeon (from Hades, the Greek god of the underworld). Life probably first appeared at the end of this aeon in a period called the Archaean (see Table 8.1 for the dating and characteristics of the different geological periods). It is conceptually of the greatest importance to remember that, however exceptional, Earth is just one of eight planets in the Solar System.

Earth began to form from the swirling maelstrom of the rocks and dust surrounding the Sun (the solar nebula) only a short time after the Sun itself coalesced into a mass with a density sufficient to drive fission activity, thus becoming a star. In relation to potentially life-supporting conditions, Earth's first 0.5–1 Gy are well expressed in the words from the Hebrew bible's story of the creation '*Tohu V'Vohu*'—complete chaos. This was a time during which there were frequent collisions with bodies varying in size from rocks to planetisimals measured in kilometres; the 'bombardment phase'. These impactors originated from the same solar nebula and accreted to form planet Earth. As it formed, Earth acquired the physical characteristics described in Chapter 4. This period probably looked much like the artist's impression shown in Fig. 5.1. The final form of Earth included oceans, an early land mass, and a swirling atmosphere.

As discussed in Chapters 3 and 4, the main properties of Earth making it suitable for to life are its composition from 'metallic' (heavier than helium) elements, a mass and gravity capable of holding an atmosphere, its favourable location within the Solar System, and liquid water on its surface. Simulation modelling of the Hadean aeon indicates that the major part of Earth accreted rapidly, within 100 million years of the formation of the Sun. After about 0.5–1 billion years the planet cooled to a temperature which allowed liquid phase water on its surface, and conditions stabilized sufficiently for the appearance of the first life. This was the generally accepted picture until the last

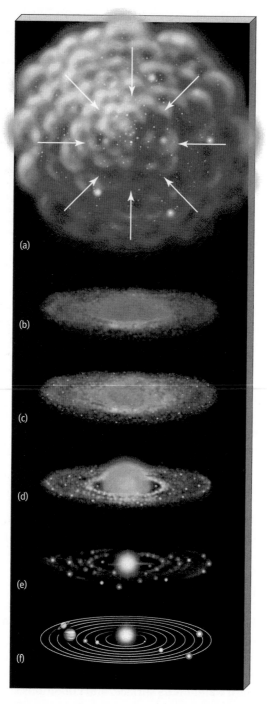

(a)

(b)

(c)

(d)

(e)

(f)

Fig. 5.1 Artist's impression of the first half billion years in the formation of the Solar System. Courtesy of NASA.

decade. Recent analysis of some of the oldest crystals surviving on Earth, zircons found in Australia and dated to 4.4 billion years ago, suggests that the Earth cooled much earlier than previously believed (Valley *et al.* 2002). There may have been primitive life present toward the end of the Hadean, but geological conditions were such as to leave no fossil or other evidence.

Origin of Earth's lone satellite, the Moon

There is some dispute as to the origin of the Moon. Today, the most commonly accepted hypothesis is that about 30–50 My after the initial formation of the Earth (4.55 Gy BP), it was impacted by a huge object having the size and mass of Mars. As a result of this collision a significant part of the original planet was ejected into space. The Earth debris and the impactor coalesced into what became the Moon. It settled into an orbit around Earth, at the relatively close distance of ~30,000 km. Analysis of rocks returned from the Moon and other evidence indicates that about 15% of the Moon was formed of material from Earth and 85% from the impactor. Unlike most of the planets, some of which have scores of moons, it became Earth's only satellite. It is, however, one of the largest planetary satellites in the Solar System, with a diameter one-quarter that of Earth, a mass 0.0123 times that of Earth, and gravity of $0.17g$. The other terrestrial planets have almost no satellites, apart from the two very small moons of Mars, Phobos and Deimos, which are less than 10 km in diameter and have negligible influence on the planet they orbit.

The Moon has almost no atmosphere, is completely dry, and, apart from some bacteria which may have survived from the astronaut visits of 1969–72, apparently bears no life. Being cool it produces no short-wave radiation of its own, but reflects short-wave from the Sun and, to a much smaller degree, from Earth. The latter, termed 'Earthshine', is the low-intensity light which can be seen on clear nights with the unaided eye on the Moon's night side. Note that the intensity of even the full Moon on a clear night is smaller by a factor of 450,000 than that of the Sun, hence its very small, usually insignificant, photo-biological effect. Having no atmosphere or eroding weather, the Moon shows the remains of comet and asteroid impacts going back almost to the time of its formation. The larger of these impact scars is what we see from Earth with the unaided eye. The significance of the Moon to life on Earth is discussed below.

Origin of Earth's water

As seen from afar, Earth is a planet of liquid water, with water covering some 70% of its surface. Life emerged in and adapted to the unique physicochemical properties of this very small, ostensibly simple, and Earth-ubiquitous molecule. Chapter 1 described the distribution of water on Earth and its supreme importance to life. Land life, from plants to advanced mammals, adapted to the low-salt water which appears on land. This 'fresh water' (containing < 0.5 g l^{-1} of salts) came from water evaporated from the oceans, followed by condensation and rain. The saline oceans ($\sim$35 g l^{-1} salts) have their own plethora of life, from microbes through photosynthesizing algae and higher plants (mainly in shallow, shore niches) to advanced mammals, such as dolphins, whose ancestors evolved on dry land.

A large part of Earth's subsurface water is trapped in minerals such as amphiboles, mica, and clay. A major question is where all the free water came from. First, as pointed out in Chapter 1, it must be remembered that although free water covers most of the planet and forms considerable subsurface aquifers, it constitutes only a small percentage of the total mass of Earth. The origin of water can be discovered from its signature—the proportion of deuterium and hydrogen isotopes (D/H) it contains. It would be expected that the D/H ratio of Earth's water (formed from the hydrogen of the nebula) would be the same as the D/H ratio in the Sun. However, the D/H ratio of the oceans is found to be very different from that in proxies of the Sun (in this case remnants of the solar wind found in lunar soil). For some time it was thought that the Earth's water came from numerous comets during the early 'bombardment' stage of Earth's formation (Delsemme 2000, 2001). However, when isotope samples were obtained from comets, the D/H ratio was found to be very different from that of ocean water. Moreover, it appears that most of the impacts of the great early bombardment were of dry asteroids rather than water-rich comets. Most of Earth's water appears to have come from the accretion of bodies from the outer asteroid belt during the late stage of Earth's formation. This is learnt from the signature of water found in carbonaceous chondrites (see Chapter 4). Only some 10% of Earth's water appears to have originated from comets from the Kuiper belt and the Uranus–Neptune region of the Solar System (Morbidelli et al. 2000). There are, however, methodological difficulties in these studies (Robert 2001) and conclusions are still uncertain. Several processes, such as condensation and evaporation, change (fractionate) the D/H ratio.

Another possible source of a considerable part of Earth's water is the reduction by early free hydrogen of the iron oxide then prevalent on Earth (FeO + H$_2$ → Fe + H$_2$O). Although the exact mechanisms of the formation and maintenance of the oceans are poorly understood, it is clear that liquid water was present on the surface of the Earth quite a short time after its formation (here meaning tens to hundreds of millions of years).

Earth is not unique in the Solar System in having significant liquid water (see, for example, the satellites Europa and Enceladus – Chapter 4). It is the early appearance and lengthy *persistence of liquid water on its surface* that is unique and has enabled the gradual evolution of advanced life. From the perspective of life, the source of the water is of secondary importance. With apologies to Gertrude Stein:[1] water is water is…!

---

[1] Gertrude Stein (1913): 'A rose is a rose is a rose is a rose is a rose'.

Origin of Earth's early and second atmospheres

What has been called Earth's 'first atmosphere' was almost certainly composed mainly of hydrogen and helium. These two gases, both very light, were soon dissipated, being lost to space due to the heat of the still forming Earth and the impact of the solar wind. After about 200 million years Earth cooled and formed a crust. Numerous volcanoes began pouring out gasses from the mantle of the primordial Earth. These included steam (about 95%), CO_2, ammonia, methane (CH_4), chlorine, sulphur, nitrogen, and hydrogen. These are the gases ejected by volcanoes to this day. (Note the absence of free oxygen.) It is assumed that the composition of volcanic ejecta has remained more or less constant. A very dense, permanent atmosphere was created in this early volcanic period, with a pressure of about 100 bars.[2] As the Earth cooled, much of the CO_2 dissolved into the oceans, with a considerable portion forming $CaCO_3$ precipitates which settled on the sea floor. Later deposits of calcium came from biotic sources (see below). The remaining atmosphere contained water vapour, CO_2, some nitrogen, and CH_4. This has been termed Earth's 'second atmosphere'. It contained little oxygen. There may have been some free hydrogen. It should be remembered that in order to hold on to an atmosphere, the Earth had to have a considerable gravitational field—which is a function of size and mass. This was attained early on, once Earth had accreted sufficient mass from other bodies of the solar nebula.

The second atmosphere was rich in the so-called greenhouse gasses. Among the most important of these were CO_2 and CH_4. In that primordial atmosphere CH_4 was only a small component compared with CO_2. Nevertheless, mole for mole, methane absorbs far more infrared radiation, making it a powerful greenhouse gas. Together, CO_2 and CH_4 kept the surface temperature at around 70°C, despite the faint early Sun. Ammonia (NH_3), a far more potent greenhouse gas than either CO_2 or CH_4, was also ejected by volcanoes, but was rapidly destroyed by UV radiation.

Kasting (2004) suggests that at this time CH_4 was the main greenhouse gas responsible for the relatively hot early climate. To have achieved the same level of absorbance of infrared radiation, CO_2 would have had to be very concentrated. At such concentrations it would have dissolved in water to form HCO_3 and CO_3, which in turn reacted with the then available Fe(II) to form the mineral siderite (mainly $FeCO_3$). Methane became so thick that it eventually polymerized into complex hydrocarbons, which formed a high-altitude haze, similar to that of Saturn's satellite Titan as seen today (Chapter 4). Similar to Titan, the CH_4 in Earth's first atmosphere was not of organic origin. The CH_4 in Earth's second atmosphere is believed to have been produced by methanogenic anaerobic microbes (Kasting and Siefert 2002; Pavlov *et al.* 2003). These microbes first appeared in the Archaean. They formed one of the first three original domains of life, the other two being the bacteria and the eukaryotes (Chapter 6).

---

[2] One bar is the present atmospheric pressure at sea level.

Earth's third, oxygen-rich, atmosphere, emerged only 1.7 billion years after the formation of Earth (i.e. ~2.8 billion years ago). It resulted from the activity of oxygenic photosynthesizing cyanobacteria, which split water and released oxygen. This momentous change in the biosphere is discussed in Chapter 6.

Earth's magnetosphere

We are all acquainted with the magnetic field around small bar magnets, the presence of which can be demonstrated by the arrangement of iron filings. An apparently similar field is found around Earth (why 'apparently' is discussed below). This field gives Earth its North and South magnetic poles (at present about 11.3° from the geographical poles, which are the tips of the Earth's spindle of rotation).

The region around the Earth affected by its magnetism is called the 'magnetosphere'. However, the shape and characteristics of the magnetosphere are determined by numerous factors, from Earth and from space, as depicted in Fig. 5.2. Note that although Earth's magnetism is relatively weak, varying from 0.3–0.6 gauss,[3] the magnetosphere is huge in comparison to the size of the Earth. It extends tens of thousands of kilometres from Earth. Its geometry is determined by the movement of Earth in space and by the solar wind. Nearer to Earth its most important characteristic is that it stretches between the magnetic poles and deflects most of the incoming cosmic radiation from the solar wind to the poles. In regions of high latitude it gives rise to the phenomenon of the aurorae (aurora borealis in northern skies, aurora australis in the south).

The bar magnet analogy is useful, but not exact. Earth's magnetic field does not surround a fixed magnet. It is believed to be induced by the electrical currents generated by the flow of Earth's liquid outer core around its solid inner core; a dynamo effect. The movements of the different layers and the induction of electric currents, which develop the magnetic fields, are highly dynamic and not completely understood. One of the more complex of these motions results in the magnetosphere reversing its direction every 100,000–250,000 years, with no apparently regular periodicity. When this happens, the North and South poles exchange positions.

Biological significance of Earth's magnetosphere

One of the most ancient human uses of the Earth's magnetism was in navigation. The 12th-century Chinese used a form of magnetized iron oxide (Fe_3O_4) called magnetite. They suspended pieces ('lodestones') from a thread, to show the North/South pole alignment—the first magnetic compasses. This was an important aid, especially under conditions in which the stars, such as Polaris, could not be seen. As described below, the positions of the magnetic poles wander. This caused considerable difficulty up to

---

[3] The SI unit for magnetic flux density is the tesla (T). One gauss (1 G) = 10^{-4} T. By way of comparison, a toy magnet made of neodymium, has a strength of the order of 10^4 G or 1 T.

Fig. 5.2 Artists impression of Earth's Magnetosphere (on the right) and the Sun with its Corona and Solar Wind (on the left). Courtesy of SOHO, ESA and NASA.

the time of GPS satellite navigation. Some species of birds can also sense the Earth's magnetic polarity and use it to guide their migrations.

It is sometimes stated that the magnetosphere protects life on Earth from cosmic ray-induced mutations (see Chapter 4 and Sobel, 2005, for example). A planet such as Mars, which lost its magnetosphere, would have too much ionizing radiation to support life. This is much over-stated. One instance in which the lack of a magnetosphere may indeed have been significant is on Mars, but for another reason. In the period before about 500 My BP Mars had an atmosphere. Lacking a magnetosphere and strong gravity, the solar wind may have been an important factor in this atmosphere having being blown away into space.

Most of the high-energy ionizing cosmic radiation reaching Earth comes from galactic sources. Some arrives from occasional (with time measured in years) high-energy solar flares. The attenuation of such radiation by Earth's magnetosphere is a minor factor for life, as most is absorbed by the upper atmosphere. Some radiation from solar flares does penetrate, however. It then affects ground-based electronics. Moreover, it can seriously damage space-borne communication satellites unprotected by the atmosphere and may affect astronauts and even the crew and passengers of high-flying commercial planes.

In Chapter 4 it was noted that the small part of the high-energy galactic cosmic radiation which does reach the biosphere, while not being generally catastrophic to life, is of considerable biological significance. This is because it is one of the important factors inducing mutations. Mutations, while usually deleterious, may very occasionally (and quite serendipitously) produce solutions to new environmental challenges, thus advancing biological evolution.

Evidence for the lack of critical importance of the magnetosphere in shielding life from cosmic radiation is provided by the periods of polar reversals. These reversals are recorded in magnetite-like rocks, and are clearly seen in those around the mid-ocean ridges. The hot magma ejected periodically at the ridge quickly solidifies into more or less linear and datable, lines of rock. These rocks record the magnetic orientation of the globe at the time of their formation. Adjacent lines are found periodically, which have opposite-facing magnetic polarity.

During periods of polar reversal the Earth's magnetic moment, and with it the magnetosphere, are much reduced. Even so, there is no palaeogeological evidence of any large-scale biological effects, such as destruction of species or a widespread increase in the rate of mutations. At such times migratory birds may have had navigation problems.

Today we are in a period when the Earth's magnetism is declining. It has weakened by 5% in the last century, since the first accurate measurements were made. It may suddenly reverse; just when is not known. For a review of planetary magnetospheres see Van Allen and Bagenal (1999).

The end of the Hadean aeon

After 1 Gy Earth became sufficiently quiescent for primitive life, although the physical properties of Earth did not cease to evolve. As described in Chapter 4, surface features (most predominantly driven by the moving tectonic plates) changed, and they continue to change to this day. The same goes for the composition of the atmosphere (see above and Chapter 6). Also, as discussed above, solar radiation was steadily increasing: about 4.5 Gy BP radiation was 30% less than it is today. Life had to contend with and adapt to these and to many other changing features of planet Earth; for example, day length. Early Earth rotated once in about 10 hours, but, under the influence of the receding Moon, slowed to once in 20 hours by the beginning of the Phanerozoic aeon (550 My BP). The speed of Earth's circumnavigation of the Sun gradually increased. In the early Phanerozoic the length of the year was about 400 days. It slowly accelerated to today's year of 365 days. As discussed in Chapters 3 and 4 and below, a slowing speed of rotation and an increasingly faster solar orbit gradually modified the climate of the biosphere in ways to which life had to adapt.

The first half of the Hadean aeon was certainly unsuitable for life, being characterized by very high temperatures and frequent impacts from the debris of the solar nebula. A review of what is known of the formation of early Earth can be found in Nisbet and

Sleep (2003). Life, which may have first appeared in the last hundreds of million years of this aeon, had to adapt to the environment available. However, then, and later during its 3.7 Gy history on Earth, life has had to contend with erratic environmental changes. Some were catastrophic (Chapter 7).

Motions of Earth in its circumnavigation of the Sun

The motions of Earth in its circumnavigation of the Sun are neither monotonic nor constant. To a large extent they determine the dynamics of climate, from day length to seasons and from ice ages to interglacial periods. Again, living species had to adapt to the changing environment or perish. It is important to remember that most advanced species have failed to survive for more than about a million years.

The present motion of the Earth around the Sun is shown in Fig. 5.3. The Earth is revolving around the solar plane (the ecliptic) at an angle; i.e. the North/South axis is now at 23.5° (Figs 5.3 and 5.4). This inclination or obliquity means that the Earth presents a different face to the Sun as the year advances. This is the basis of climate and seasons. In the Northern Hemisphere winter the northern half of Earth is slanted away from the Sun, which reduces the length of the day and the intensity of the sunlight reaching the surface. The reverse is the case in summer. The process is the same, but reversed again, in the Southern Hemisphere. In the Northern Hemisphere the longest day is on 21 June and the shortest on 21 December, again with the reverse in the Southern Hemisphere. The seasons are also affected by the distance from the Sun, which changes in the course of the year (see below).

Fig. 5.3 Present-day motion of the Earth in its circumnavigation of the Sun. Not to scale. Courtesy of Wikipedia.

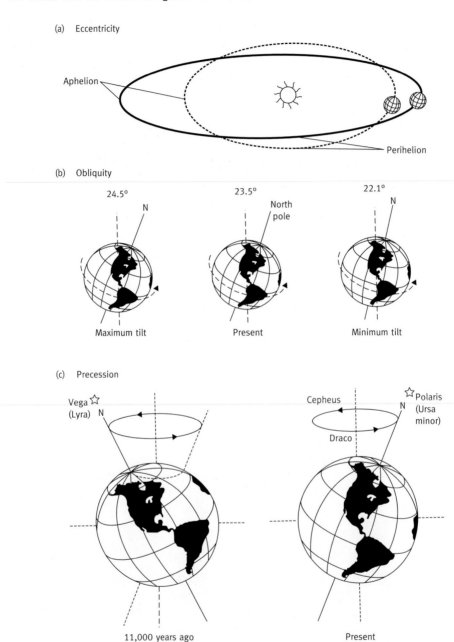

Fig. 5.4 Motions of the Earth during its annual circumnavigation of the Sun. Eccentricity, the deviation from a true circle towards an elliptical orbit, results in the Earth being closest to the Sun (perihelion) during the Northern Hemisphere's winter. Obliquity is the angle of Earth's rotation in relation to the ecliptic, the plane of its solar circumnavigation; it changes the presentation of the surfaces of the Earth toward the Sun, resulting in seasons. Precession changes the polar positions of the Earth in a circular movement (this used to require corrections in star-based navigation). Redrawn from Gates (1993).

There is of course little seasonal change in the radiation incident on the tropical regions near the equator. Where there are seasonal changes in climate, e.g. the monsoons on the Indian subcontinent, they are a result of world-wide atmospheric circulation patterns and the interplay between different heat capacities of the local land and ocean masses. When the land heats up faster than the sea, air over the land rises and hot humid air flows in from the oceans, producing heavy seasonal rains.

The obliquity of other planets in the Solar System is often very different from that of Earth. Consider, for example, Uranus, whose obliquity is about 90°. This causes its southern pole to constantly face the Sun. There are no changing seasons on Uranus. Assuming that it was within the CHZ (which it is not) life could possibly adapt to such conditions in some regions of the planet. There would be bands of suitable temperature between its very hot south pole and its freezing north pole, where liquid water could exist. Any life which evolved under these conditions would be very different from that on Earth, in ways difficult to foresee (science fiction authors may apply their imaginations!).

As discussed in Chapter 4, the first and perhaps most biologically significant movement of Earth, one which we tend to take for granted, is its rotation—at present once every 24 hours. Since the formation of Earth 4.55 Gy BP the day has become progressively longer from its initial 10 hours. There is much variance in the motions of the planets (Table 5.1). Rotation is not a necessary characteristic. A planet could rotate with the same side constantly facing its star, just as the Moon has one side constantly facing Earth.

The length of the year also varies. As explained by the physical laws of planetary motion developed by Newton and Kepler, the further it is from the Sun the longer it takes for a planet to complete one orbit (by definition that planet's year). Even our neighbour Mars has a year almost twice as long as that of Earth. Our other neighbour Venus, nearer to the Sun, has an orbit of just 225 days but a day length of 243 Earth days. Again, assuming that other conditions were suitable for life, these rotation and orbiting speeds would produce climates very different from that on Earth.

Table 5.1 Rotation and periods of the solar planets

Planet	Rotation speed	Orbital period (Earth days)
Mercury	58.6 Earth days	88
Venus	243 Earth days	224.7
Earth	1 day	365.23
Mars	1 Earth day	689
Jupiter	10 hours	4332
Saturn	10 hours	10,759
Uranus	17 hours	30,684
Neptune	16 hours	60,190

In its journey around the Sun the Earth exhibits three irregularities. The first is obliquity, discussed above. The angle of obliquity is not constant but varies between 22.1° and 24.5° over a period of 41,000 y. It is currently at 23.5°. The variation does not seem very great, but it changes the geographical distribution of solar energy and hence climate. The other two irregular motions are eccentricity and precession. These three movements of the Earth are shown diagrammatically in Fig. 5.4.

Eccentricity is the degree to which the orbit of the Earth is elliptical. The orbit of Earth must be an ellipse as it shares a common virtual centre of gravity with the Sun. As the Earth has only a very small gravity compared with the Sun this ellipse shows only a small but not insignificant departure from a true circle. At present, in the Northern Hemisphere, Earth is nearer to the Sun (at perihelion) in winter than in summer (at aphelion) (Fig. 5.4). Consequently in the north there is about 7% more solar radiation incident at the top of the atmosphere in mid-winter (January) than in mid-summer (July). This moderates the climate of the Northern Hemisphere. In the Southern Hemisphere this radiation effect is reversed, and winters and summers are potentially more extreme. 'Potentially', as there are many factors determining climate other than incident solar radiation. Like all the movements of Earth, eccentricity is not constant. It varies between 0.01 and 0.07, with a periodicity of about 96,000 y. Both obliquity and eccentricity are affected by the gravitational pull of the other planets, especially Jupiter and Saturn.

As shown in Fig. 5.4, the axis of Earth's rotation tends to wander in a circle with a periodicity of 22,000 y. This movement is called precession. It is caused by the gravitational pull of the Sun and Moon. This is a relatively short cycle and results in a periodic change in the timing of the winter and summer solstices,[4] currently 21 December and 21 June, respectively (Fig. 5.3). Eleven thousand years ago the summer solstice (the longest day), in the Northern Hemisphere, would have fallen on 21 December. These changes may have had little influence on natural vegetation or fauna but, in a historical setting of millennia, calendars used by farmers to determine the seasons of planting and harvesting would have been found to slowly wander between the seasons. For example, if the sowing of wheat in Babylon in the ancient Middle East about 2000 BCE was scheduled for early February, then, by around 0 CE, it would have moved to mid-April. (The calendar and names of the months were of course quite different in ancient Babylon.)

In the days before GPS, navigators often gauged their position from the stars, especially Polaris, the 'North Star', but because of precession, 11,000 y BP the North Pole pointed toward Vega (Fig. 5.4).

In 1920, Milutin Milankovic, a Serbian mathematician, published his theory of ice ages. According to this, the combination of the above three Earth irregularities (eccentricity, obliquity, and precession) results in reduced insolation incident on Earth with a

---

[4] Solstices are the longest and shortest days of the year. In the Northern Hemisphere they fall between 20–21 July and 21–22 December in summer and winter, respectively.

periodicity which correlates with the appearance of ice ages. Although there were many exceptions, especially in the much longer and more ancient of the ice age periods, his model seems to work quite well for the ice ages of the Quaternary period (the last 1.6 My). There are, however, other theories explaining the induction of ice ages (see Chapter 7).

Significance of the Moon to life on Earth

> *The moon is nothing but a circumambulating aphrodisiac, divinely subsidised to provoke the world into a rising birth-rate.*
>
> Christopher Fry (1948) The lady's not for burning

Most quotations citing the Moon are more lyrical than the above. However, in addition to inspiring poetry and humour, and providing some illumination on clear moonlit nights, the Moon is of surprising biological significance.

Despite its bleak physical characteristics, the Moon was of considerable importance to the appearance of life on earth 3.7 G BP and its subsequent evolution. Indeed many authors (e.g. Ward and Brownlee 2000) believe that its very existence was and is absolutely essential for life. Comins (1993) devoted an entire book to explaining how small deviations from just one of the physical characteristics of the Moon would have presented severe and perhaps insurmountable problems for life on Earth.

The Moon affects life in many indirect and very occasionally direct ways, the following being the more outstanding indirect effects:

- Day length: as discussed above, when the early Moon was close to Earth, the Earth was spinning much faster than today, with a day being only 10 hours. As tidal friction accrued the Moon spiralled outward and the rotation of the Earth slowed to what it is now, a 24 hour day.
- Tides: the Moon and the Sun produce oceanic (and atmospheric) tides on Earth as a result of their gravitational pull. When the early Moon was close to Earth it was the dominating force and tides were then huge and cycled every few hours. According to one rather speculative theory (Lathe 2004) the fast tidal wetting and drying cycles (every 2–6 hours) could have induced the necessarily repetitive conditions required by polymerase reactions which may have produced the first DNA-like molecules. Each step in polymerization requires dehydration followed by rehydration. This could have been an essential step in the appearance of life (Chapter 6).
- Seasons: the Moon stabilizes Earth's obliquity (Fig. 5.4). As explained above, obliquity is the cause of seasonality as it presents the different regions of Earth toward the Sun at varying but recurrent angles over the period (year) of its orbit. Laskar *et al.* (1993) calculated that without the gravitational stabilization provided by the presence of the Moon, the obliquity would change frequently with no regularly recurrent pattern. It would have been very difficult for advanced land-based life to cope with the resulting

irregularly changing climatic conditions. An extreme change in obliquity could cause runaway ice ages, with alternating warm periods (Chapter 7). Mars has about the same rate of spin as Earth but an obliquity (currently 45°) which changes frequently. Without a single, massive, stabilizing satellite such as Earth's Moon, the seasons during Mars' putative wet period would have been chaotic. Contrariwise, Mercury, the closest planet to the Sun, has almost no tilt (i.e. an obliquity of zero) and no satellites. As a result, while most of its surface is roasted by the Sun its poles are cold enough to support ice.

- Tectonics: a number of authors (e.g. Ward and Brownlee 2000) have suggested that the violent formation of the Moon may have engendered Earth's plate tectonics. As discussed above, tectonic activity is very important and perhaps vital for life on Earth. Moreover, what seems to be an unsubstantiated but interesting conjecture holds that without the initial removal of a large part of the Earth's crust by the impactor which produced the Moon, the ocean level would have been higher. If so, there would have been very little room for the evolution of dry land flora and fauna (see http://www.spacedaily.com/news/life-01x2.html).

The direct biological effects of moonlight on plants, insects, and animals, are generally of secondary importance. But there are a few on record:

- Vision: some animals, including humans, have eyes which are sensitive enough to utilize the light of the moon. Armies use moonlight (or did, until the advent of infrared night vision equipment). Insects which are hunted by nocturnal enemies tend to avoid moonlit nights. This is anecdotal. Reliable data are hard to find. For the same reason farmers spraying insect pests may choose dark nights.
- Timing and morphogenesis: the involvement of moonlight in a biological process, that is supported by more data, is the circadian response of wheat. In wheat, moonlight has been reported to be sufficient to invoke a response of the ubiquitous plant light switch, phytochrome, involved in circadian rhythms (Nagy *et al.* 1993). Phytochrome is involved in numerous plant responses (Chapter 3), so moonlight may have other morphogenetic effects on plants. Moonlight is certainly too low in intensity to drive photosynthesis.
- Sex (I knew I could get this into the text somewhere!): one surpising effect of moonlight on sexual activity in animals has recently been reported for the lowly corals. Coral reefs, which are groups of normally sedentary animals, have long been observed to exhibit an extraordinary synchronous spawning. After as long as a year with no sexual activity, and within a period of days, innumerable eggs and sperm are released simultaneously into the sea by millions of individual corals, engendering an orgy of sexual reproduction. It has recently been found that the trigger is moonlight, absorbed by a blue light-sensitive cryptochrome pigment of the corals during periods of full moon (Levy *et al.* 2007). So, maybe the playwright Christopher Fry, quoted above, knew some biology!

Jupiter—Earth's guardian angel (perhaps)

Although the Earth became more tranquil after the first billion years of the Hadean aeon, it did not become a 'rose garden'. Throughout the 3.7 Gy of life's evolution it was challenged by a forever changing, sometimes catastrophic, environment. This included an atmosphere which became toxic to many life forms and has remained partly toxic to plants and animals (Chapter 6). Life has had to contend with ice ages, mega-volcanoes, eruptions of huge quantities of methane or hydrogen sulphide from the seas, and occasional impacts by still roaming remnants of the inner solar nebula. In addition to these, there have been collisions with wandering comets and errant asteroids from the Oort cloud and from the main and Kuiper asteroid belts. These recurring calamities are described in detail in Chapter 7. Here we take a closer look at Jupiter, the largest planet in the Solar System, which *may* (read on) have played a special role in stabilizing conditions on Earth and reducing the frequency of comet and bolide impacts.[5]

Jupiter is the fifth planet from the Sun. Unlike the inner terrestrial four, it is composed mainly of gas. It does not have any defined surface. In this it is similar to the other three gas planets of the Solar System—Saturn, Uranus, and Neptune (sometimes called the 'Jovian' planets). Jupiter is composed mainly of hydrogen, with some helium. In this it is very similar to the Sun. The centre of Jupiter is under enormous pressure. It may have a rocky core of heavier elements. Jupiter rotates once every 10 hours and its liquid interior swirls violently. These motions give rise to a magnetic field 20,000 times stronger than that of Earth.

Jupiter's mass is 318 times, and its gravity 2.6 times, that of Earth. Its outer layers are characterized by swirling storms, some as large as the Earth. As a result of its size and relatively low density its gravity is perhaps less than what may have been expected. Its size, mass, and gravity are what have made it so vital to the inner Solar System and to the Earth in particular. Apart from Mercury, the nearest planet to the Sun, the orbital planes of the planets are actually closer to that of Jupiter than to the Sun. Jupiter is responsible for the relative orbital stability of the four inner, terrestrial planets. It also caused many of the remnants of the inner primordial solar nebula to enter more or less permanent orbits within the main asteroid belt. This stabilized the inner Solar System within 1 billion years of its formation. Again (and again...) it must be emphasized that we have only one example of life but, from what we know of its evolution on Earth, an early beginning was important. It took all of 3.5 Gy for evolution to go from early Archaean life to *Homo sapiens sapiens*, under *relatively* stable conditions. As noted above, this is a quarter of the entire lifetime of the universe since the Big Bang, 13.7 Gy BP.

As discussed above and in detail in Chapter 7, Earth has been repeatedly struck by comets and asteroids, ranging in size from a few grams to thousands of tonnes. The larger impacts have resulted in environmental disasters from a local to a global scale.

---

[5] A bolide is a large meteorite, usually one that explodes.

Because of the tremendous gravitational attraction of Jupiter, many of the potential impactors have ended up striking Jupiter, or have been diverted into orbits which spare Earth. In 1994, for the first time, a wandering comet, named after its discoverers Shoemaker-Levy 9, was seen to fragment and impact with Jupiter. Even without Jupiter it would not have struck Earth, but the sight was amazing and, for humans, very humbling! (For a movie on this dramatic occurrence see: http://www.youtube.com/watch?v=tbhT6KbHvZ8&feature=related)

The Cretaceous period was brought to an end by the impact of a large body (a comet or bolide). This Cretaceous/Tertiary event (called the K/T; 'K' to differentiate it from the Carboniferous period) occurred 65 My BP. The object fell in the Yucatan Peninsula of what is now Mexico. It is perhaps the one extraterrestrial impact event for which there is solid evidence connecting it to an environmental disaster of global dimensions. The temporary change in climate it brought about destroyed most life in the biosphere. Most famously it ended the age of the dinosaurs. There have been at least five massive global extinction events during life's evolutionary history, and numerous smaller-scale events. Although there are many suggestive indications, there is no firm evidence that extinction events of a world-wide scale, other than at the K/T boundary, have been a result of extraterrestrial causes (Hallam and Wignall 1997; Courtillot 1999; Hallam 2004; see also Chapter 7). On the other hand there have been numerous confirmed instances of impacts of extraterrestrial origin which probably resulted in severe local disruption of life (Chapter 7).

If major impacts had occurred more frequently, it is doubtful whether advanced life could have evolved. This is where Jupiter our 'guardian angel' comes into play. According to Wetherill (1995), in the absence of Jupiter Earth would be impacted at a frequency about a thousand times greater. Assuming for a moment that events with a magnitude of the K/T impact occur once in 65 million years then, without Jupiter, they would occur once in 65 M/1000, i.e. every 65,000 years. Under such a bombardment *Homo sapiens sapiens* would certainly not have appeared and persisted, even for the present 200,000 years (Ward and Brownlee 2000). However, note that the title of this section, presenting Jupiter's 'guardian angel' effect, had the caveat 'perhaps'. This is because recent calculations by Laakso *et al.* (2006) indicate that the protective effect of Jupiter may be much smaller than previously thought, or even entirely lacking. Others have calculated that if there is a protective effect of Jupiter it would be mainly for the long-term comets originating from the Oort cloud and much less so for the short term comets from the Kuiper belt, or for asteroids. The jury is still out!

In summary, the presence of a single, massive satellite, our Moon and of a nearby vastly more massive planet, Jupiter (but bearing in mind Laakso *et al.*'s reservation) were and are of enormous importance to life on Earth. They are just two of the very many essential, but purely fortuitous factors which *together* produced the biosphere: the thin layer of the planet which had and still has the environmental conditions which enabled the emergence, evolution, and persistence of life.

Summary

- During the first 1 Gy of Earth's existence, the Hadean aeon, Earth accreted from the solar nebula. During this period Earth was frequently struck by asteroids and comets. Very early in the Hadean, Earth was impacted by a body the size of Mars, which tore off part of the Earth. The impactor and part of Earth broke away, forming the Moon.
- Most of Earth's water came from within the Earth, from the reduction of iron oxides by hydrogen. A small fraction was brought to Earth by impacting comets.
- Earth's initial atmosphere was composed mainly of hydrogen and helium. It was lost to space.
- Earth's second atmosphere came from outgassing of volcanoes. It was composed of about 95% water vapour with CO_2, ammonia, and CH_4 constituting the remaining 5%. It contained little oxygen. Most of the ammonia molecules were broken down by UV radiation. Much of the CO_2 reacted with calcium minerals and precipitated into the oceans. The remaining gases exerted sufficient greenhouse effect to keep the planet's surface warm, despite the low intensity of the early Sun, which was then about 25% less than it is today. Life began in this second atmosphere.
- The flow of Earth's liquid outer core around a solid inner core produces an electrical charge which induces a magnetic field surrounding the Earth—the magnetosphere. This slowed the dissipation of Earth's early atmosphere by the solar wind. Today it is of little biological significance but does affect electronic communications.
- In its circumnavigation of the Sun, Earth shows four secondary movements of significance to life: rotation, which produces day and night; obliquity, or angle to the ecliptic, which results in seasons; eccentricity, the degree to which its passage around the Sun is elliptical, which modifies climate; and precession, the wobble of its rotational axis, which changes the timing of the solstices. All four movements change periodically, together and individually affecting the climate of the biosphere.
- The presence of a single massive satellite, the Moon, tends to stabilize Earth's movements and hence its climate. This has been a very important enabling factor in the evolution of life.
- Earth's movements are affected, to a small extent, by the gravitational pull of the other planets. The 'nearby' presence of a very massive planet, Jupiter (which tends to stabilize the entire inner Solar System), together with the presence of the Moon, tends to dampen their influence.
- The gravitational attraction of Jupiter has been calculated to reduce the frequency of comet and bolide impacts on Earth (such as that at the K/T boundary, 65 My BP, which produced a global biological catastrophe) by a factor of a thousand. However, this factor of a thousand has been challenged. It may be far too large or non-existent.

6

Origin of life and photosynthesis

What do we know, what can we know of the origin of life?

Origins have a special place among conundrums we are all drawn to. Where did I come from? How did the universe start? How did life begin?...Yet questions about origins pose special problems for scientists...novelty puzzles us. That genuinely new things come to exist as a result of natural processes seems almost a paradox, for is not everything made up of elementary, unchanging particles? How can it be that ordinary molecules, completely describable in the language of quantum theory and chemistry, suddenly assemble themselves into living cells that can only be understood in a whole new language—that of biology?

Lee Smolin, in a New York Times *review of Paul Davies' (2000) book on the origin of life*

The first chapters of this book described how conditions favourable to life were established on the surface of Earth, in a thin unstable layer we call the biosphere. But, and it is a very big 'but', having the conditions for life does not mean that life will surely appear. Before the 19th century many believed that even advanced life forms, such as flies and mice, would form spontaneously given a suitable substrate (such as a piece of rotten meat) and just a little time (days). This idea was finally put to rest by the classic demonstrations of Louis Pasteur. 'Finally' as his was a follow-up to the work of Francesco Redi and Lazaro Spallanzani, in the 17th and 18th centuries, respectively.

The problem remained as to how life first appeared on Earth. Even the most elementary forms of life are enormously complex. A few, non-independent entities,[1] such as viruses, have just a thin protein covering. Simple bacteria and all more advanced life forms have sophisticated membranes for separating their internal milieu from the external world, mechanisms for handling energy, complex chemicals that serve as substrates and regulators of the cellular chemical processes (metabolism), tissue construction, and proteins for catalysing biochemical reactions and sometimes for special tissue formation. They also reproduce and have the means for storing data on their construction and operation and for transferring this information from one generation to the next.

---

[1] Part of their life must, by necessity, be passed in the cells of more advanced life forms.

At first sight it seems to be almost a statistical impossibility that simple life forms, or even groups of reproducing organic molecules, could have appeared spontaneously in the early 'soup'. There are two conceptual ways to avoid this problem of chance. The first is the religious solution, 'God created...', and the second 'Life came from some-where else, outside of Earth'—the Panspermia hypothesis. Both are evasive. How was God created? How did life originate on some other celestial body? The first idea is either not accepted or is actively rejected by most of today's scientists. It is unprovable or, to be more logically rigorous, not disprovable. Many (but by no means all) of the propo-nents of the theist solution believe[2] that first life was created some 5,700 y BP. Although innumerable findings of science show this not to be so, proponents of 'creationism' or, in its latest incarnation 'intelligent design', are not receptive of normative evidence and logical arguments. Contrariwise, the Panspermia hypothesis, while today having few proponents, is at least possible. Many scientists believe,[2] as did the great French biolo-gist and Nobel prize-winner Jacques Monod (1971), that given the right conditions for life *and enough time*, elementary life will inevitably and spontaneously appear (but not immediately flies and mice!). This is the basis for the optimism of astrobiologists. They seek out extraterrestrial life by first searching for environmental conditions similar to those on Earth, on other planets and planetary satellites (Chapter 10).

There is indeed a major problem in our understanding of the rapid appearance of life on Earth. As usual in science, when the answers are not known innumerable stud-ies are made and books written. There are so many books on the origin of life that I can mention only a few, no doubt omitting some equally excellent works. Among them are those by Brack (1998), Davies (2000), de Duve (1991), Dyson (1999), Fenchel (2002), Fry (2000), Hazen (2005), Knoll (2003), Lahav (1999), Schopf (2002), and Schulze-Makuch and Irwin (2006). One of the most prominent researchers of the origin of life was Manfred Eigen. A brief summary of his ideas is given in Eigen (1994). No list of books on the origin of life should fail to mention the seminal work of Oparin, first published in Russian in the 1920s. There is an English translation of the 1938 edition (Oparin 2003). Oparin's ideas were similar to those of the Englishman J. B. S. Haldane, who worked quite independently in the same period. Their work was extraordinarily prescient; their basic understanding was not very different from that of today. There are very many ongoing research projects and societies devoted to the study of the origin of life, all of which make the above books and of course this one, not entirely up to date. Many of the latest reports can be found in the journal *Origins of life and evolution of biospheres*. The Wikipedia article 'Origin of life', is particularly comprehensive and up-to-date (see References and Resources).

One of the most lucid semi-popular books on the origin of life is that of Paul Davies, a physicist and prolific author (Davies 2000). The opening quotation to this chapter is

---

[2] When religious people 'believe', they do so because they want to. No proof is required; no disproof is entertained. When scientists use the word 'belief' they are saying that they think that findings to date support a certain hypothesis—but that this may change if and when new facts emerge.

from a very positive review. Most scientists would be unhappy with its title: *The fifth miracle; the search for the origin and meaning of life.* In his preface, Davies gives a rather weak literary excuse, with biblical reference, for the 'miracle'. 'Meaning' (an irrelevancy in science) is addressed only tangentially in the final chapter. Title apart, it presents a comprehensive analysis of the scientific and philosophical aspects of the various working hypotheses on the origin of life: it is indeed a 'wonder-full' book.

As a frame of reference for discussing the various theories of the origin of life on Earth, it is important to bear in mind the time scale of some major events in the formation of Earth and the evolution of life (Table 6.1).

The data shown in Table 6.1 draw attention to a number of events, the timing of which is critical for our understanding of the origin and evolution of life on Earth. As noted, the present universe is thought to have begun its physical evolution 13.7 Gy BP. A stabilized Earth, with its guardian Moon, atmosphere, and oceans evolved 9.0–9.5 Gy later, from the debris, or nebula, of the evolving Solar System.

The first Hadean aeon of Earth, which lasted about 0.7 Gy, was unfit for life as we know it. It was too hot for liquid water and Earth was constantly impacted by asteroids and comets. At the end of the bombardment phase life appeared relatively rapidly. It was of a type now called the Archaea. The Archaea are one of the three domains of life, the others being the Bacteria and the Eukarya (which includes all advanced life). Archaea are special in having one circular chromosome and up to 30% of their genes in plasmids (extra-chromosomal genes). The Archaea contain a number of interesting groups, some significant for the formation of Earth's atmosphere. These include

Table 6.1 Major events in the history of the universe and Earth in relation to the origin and evolution of life

Time (Gy BP)	Event
13.7	The Big Bang, initiating the present universe
5.0	Formation of the Sun and solar nebula
4.57	Earth accretes from the solar nebula
4.57–3.8	The Hadean aeon. Earth frequently impacted by asteroids and comets
4.53	Earth struck by large bolide. The ejecta form the Moon
4.1	Earth's crust solidifies; oceans and atmosphere form
4.0–2.5	The Archean aeon—first appearance of life
2.5–0.54	The Proterozoic aeon—first photosynthesizing bacteria and, possibly, annelid worms
0.54	Phanerozoic aeon—evolution of multicellular organisms; many on dry land
0.25	Permian extinction event; the largest ever
0.065	Last major, world-wide extinction event
0.005	Appearance of first humanoids
0.0001–0.0002 (100,000–200,000 y BP)	First intelligent humans (*Homo sapiens sapiens*) (cynics say that this event has not yet occurred!)

the anaerobic methanogens, which produce methane, a very potent greenhouse gas. A second group, of great physiological interest, are the halobacteria. They use light energy to produce energy-rich compounds but do not have the electron transport chains characteristic of later photosynthetic organisms (see below). They do have light-activated pumps which produce ion gradients, with which adenosine triphosphate (ATP, a ubiquitous metabolic energy currency) is generated.

Apart from these special branches of early metabolism there was little obvious evolution during the first 1.5 Gy of the Archean era—'obvious' meaning just that, i.e. observable changes in the fossil and isotopic signs of its existence which survive to this day. A major advance in the evolution of life occurred 2.5 Gy BP. This was the appearance of the oxygenic photosynthesizing bacteria. Their activity radically changed the entire biosphere. The mechanism and consequences of this revolution are discussed below and in Chapter 7.

Only unicellular life existed for the next 2 Gy. Then, about 0.5 Gy BP, facilitated by the appearance of an oxygen-rich atmosphere (see below), multicellular life appeared. This brought about the final (relatively short, in terms of geological time) Phanerozoic aeon, in which we live today. It began with what has been called the Cambrian explosion. This was the sudden appearance of multicellular plants and animals which began the Cambrian period. The term 'explosion' has been debated ever since Darwin. Recently it has become more popular as the latest palaeontological studies indicate that the move from unicellular to multicellular life may indeed have been very rapid. Note in Table 6.1 that annelids are placed tentatively in the Proterozoic era. Even so, in terms of the time scale shown in Table 6.1, the Cambrian explosion, a tremendously significant progression, was very rapid. The evolution of higher plants and animals began at that time. There is still no understanding or agreement as to what caused this advance after some 4 Gy of only unicellular life.

As shown in Table 6.1, the appearance of modern humans (*Homo sapiens sapiens*) is a *very* recent event, having occurred about 200,000 y BP. This chance happening and its biological consequences for the biosphere are discussed in Chapter 9. Note that it took about 4 Gy, or one-third of the entire existence of the universe, for humans to appear. They may have evolved sooner but, on a geological time scale, not much sooner. This is because a large brain could come only after the evolution of a complex neurological system. The long period of time before the appearance of human intelligence is of great significance when we try to estimate the chances of there being other advanced living entities in the universe. Again, 'advanced life' means of a type with which we are familiar. There may be intelligent life elsewhere, which developed on a time scale quite different from our own (Chapter 10).

Panspermia

The panspermia hypothesis should be revisited before considering present ideas on how life evolved on Earth. According to this concept, life came to Earth from elsewhere in the Solar System, or even from the Milky Way or cosmos at large (see Chapter 9 of Davies, 2000, for a historical and critical review and Warmflash and Weiss, 2005, for a further update).

As noted above, the motivation for panspermia is that life forms are found in strata dated to very soon after the end of the Hadean aeon, during which life was not possible. Liquid water and oceans may have formed in the early Hadean but Earth (together with all the other inner planets and satellites) suffered a second period of heavy asteroid and or comet bombardment in the late Hadean.[3] Moreover, the fossil life forms found were already extremely complicated constructs. Their appearance by chance alone, in so short a time frame, seems at first sight to be statistically impossible, or highly unlikely. There were probably even earlier, simpler life forms, but they left no fossil evidence.

In the 19th century it was proposed that comets or meteors wandering space may graze a life-inhabited planet, pick up spores or seeds, and bear them to uninhabited sites such as the pristine Earth. In the early 20th century this idea was taken up by the Swedish chemist Svante Arrhenius, who believed that discrete bacterial spores may survive space, through which they would wander impelled by starlight and the solar wind. He was probably the first to give this hypothesis the name 'Panspermia' (Arrhenius 1908). A few others have strongly promoted this idea, going so far as to ascribe the outbreak of some planet-wide plagues to the arrival of errant pathological spores from space (Hoyle and Wickramasinghe 1979). A major problem with panspermia is that most wandering seeds of life would have to survive aeons of travel, during which they would be exposed to the vacuum, dehydration, cold, and radiation of space. However, if life originated on some other body of the Solar System, such as Mars, then it is possible that the journey to Earth was relatively short (years). Moreover, laboratory and spaceborne experiments have shown that at least a few (less than 0.1%) bacteria can survive space conditions (Horneck 2003; Warmflash and Weiss, 2005). The small survival rate obtained experimentally may be irrelevant, as only a very few microbes would be sufficient to seed life. Of particular interest is the bacterium *Micrococcus radiophilus*, that has been found alive within nuclear reactors. It has a mechanism which rapidly repairs radiation-damaged DNA. It has been suggested that this bacterium arrived from space, where its ability to survive radiation evolved over a long period of time.

It is of course possible that seeds of life survive deep within large space-travelling bodies, such as meteors and comets. There they would be protected from radiation, if not from cold and vacuum conditions. Some 300 tonnes of rocky asteroids and comets reach Earth every year, some of them quite large. In 1997, the large comet Hale-Bopp passed close to Earth. This allowed unprecedented viewing and spectral analysis of its

---

[3] The 'late heavy bombardment', between 4.1 and 3.8 Gy BP, would have destroyed any life which had appeared on Earth. Evidence of this period, which affected all the inner planets and satellites, can be seen in the craters of Mars and the Moon. On Earth such craters would have been erased by plate tectonics and weathering.

contents. In 2004 Stardust, an automated probe, collected samples of material from another comet, Wild-2, which were returned to Earth and analysed. Numerous simple organic molecules were detected in both comets, some of a type unknown on Earth. None of these molecules was as complex as amino acids. Any advanced organic molecules or life forms present within such carriers must also survive the entry into Earth's atmosphere without burning up or being destroyed on impact—no small requirement.

Horneck *et al.* (2008) subjected certain bacteria, cyanobacteria, and lichens that are known to be especially stress resistant to the pressure forces which would be created should such meteorites fall to Earth. They found that all could survive a considerable impact force. Moreover, they calculated that microbe containing fragments would be thrown back to space where they could also survive. This offers much support to the concept of Lithopanspermia (panspermia within rocks).

Much excitement was caused in the 19th century by the Orguiel meteorite, mentioned in Chapter 4, which fell in France and was studied by, among others, Louis Pasteur, who searched for signs of life. In 1969 another large meteorite, the Murchison, came to Earth in southern Australia. Both these meteorites are of the carbonaceous chondrite type which are rich in organics. The Orguiel meteorite contained the amino acids glycine and beta-alanine and is thought to be of cometary origin. The Murchison meteorite contained some 70 types of amino acids, many of which are used by life on Earth. It is thought to have come from an asteroid. Of great interest was the finding that most of the amino acids were of left-handed chirality, as exclusively used by life on Earth (see Chapter 1). Even so, no life forms, extant or fossil, were found in either meteorite. As discussed in Chapter 1, a meteorite found in the Arctic that originated from Mars *may* show signs of fossil life forms (see Fig. 1.4).

The arrival to Earth of the seeds of life from outside the Solar System is thought to be extremely improbable, not to mention from outside the Milky Way. The reason for this is the huge distances between star systems. This means a low seeding density, low probability of arrival at a suitable target and long transit times.

It should be recognized that within the Solar System, panspermia, if it does occur, would be a two-way process. Seeds of life could just as well have gone from Earth to other planets and satellites as the reverse, although in the case of Mars conditions amenable to life may have evolved before those on Earth.

The empirical 'bottom-up' approach to understanding the 'rapid' appearance of life. Can we produce life in a test tube?

Before the 19th century, organic molecules were thought to be imbued with a peculiar and undefined 'vital force'. This concept was overturned in 1828, when the chemist Friedrich Wohler synthesized urea ($CO(NH_2)_2$), an organic component of urine, from the inorganic molecule ammonium cyanate (NH_4OCN).

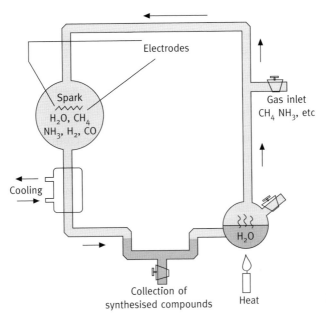

Fig. 6.1 The Miller–Urey (1953, 1959) experiment. Newly synthesised organic compounds are collected from the lower faucet.

In the 1950s Miller (1953) and Miller and Urey (1959) published the results of what has become a classic experiment. They reported the formation of organic from inorganic molecules in a 'soup' of liquids and gases. Their mixture simulated the putative conditions prevailing during the period of the 'second atmosphere' (see Chapter 5). It contained water (H_2O), methane (CH_4), ammonia (NH_3), hydrogen (H_2), and carbon monoxide (CO). Electrical discharges, simulating lightning, were passed through this brew (Fig. 6.1). At the end of a week they found that about 15% of the carbon had been converted into carbon compounds. The most significant of these were 13 of the 22 amino acids which are the backbone of proteins. These results were criticized for a number of reasons. Later studies suggested that the primordial atmosphere was not reducing. It was argued that had it contained some oxygen there would have been no synthesis of organic acids. Moreover, Miller and Urey obtained only a racemic mixture (one having equal amounts of L- and D-forms) of amino acids, instead of the strictly L-form amino acids found in all of Earth's life forms (Chapter 1). It should be noted that although most biomaterials are made from L-form amino acids, D-amino acids are frequently found in bacterial cell walls, some antibiotics, and elsewhere. Moreover, more recent studies suggest that the second atmosphere may indeed have been reducing (Tian *et al.* 2005). It is interesting to note that Oparin demonstrated that organic molecules could not be synthesized in an oxidizing milieu. From this he deduced that that the early atmosphere must have been reducing. Studies on evolution at the molecular level suggest that first life forms may have been formed from the very 13 amino acids produced in the experiments of Miller and Urey (Brooks *et al.* 2002). It is also possible that L-chiral amino acids were selected from the racemic mixture. It must, however, be realized that the number of steps from the formation of bricks (e.g. amino acids) to the building of a house (a living cell) is vast.

Although there have been many replications and modifications of Miller and Urey's work, in which other organic molecules were produced, no life forms, however simple, have appeared.

Despite the chance formation of a *reproducing* organic compound being very small, the original milieu in which this could have taken place was enormous. Miller and Urey, and the researchers who repeated and expanded their work, searched for the appearance of organic molecules and signs of life in at most a few litres of liquid; but the 'primeval soup' was the 1.38×10^6 km^3 (1.38×10^{18} litres) of the oceans.[4] In such a volume the chance appearance of a self-reproducing molecular group would not have been so small. Moreover, such a basic reproducing unit would have rapidly spread throughout the oceans, wherever there were favourable physicochemical conditions. This is because there were *absolutely no natural (living) competitors*. Consider a single bacterium of any common species, introduced into a sterile Petri dish with agar as a substrate. It soon multiplies exponentially and covers the entire surface. In the primeval ocean only local extremes of temperature or salt concentration (or composition) would prevent such a hypothetical molecular group from reproducing and spreading throughout the ocean in just a few tens or thousands of years. The many billion copies of such a replicating molecular group would then have been ripe for Darwinian evolution. Charles Darwin expressed similar ideas in his correspondence, as quoted in the Wikipedia article 'Origin of life' (see References and Resources).

Before reviewing some of the present ideas on the origin of life there is another problem to be considered. How many extant species of life are there whose origins we must explain? A first look at Earth shows it to be teeming with millions of life forms (Table 6.2a). A closer look reveals the data point in Table 6.2(b).

As shown in Table 6.2(a) there are indeed millions of life forms. But, as presented in Table 6.2(b), accurately if a little dramatically, to date only one type of life has been found on Earth. Every single species of life so far studied is built along the same basic plan. For example, molecules such as DNA, amino acids, and sugars have the same chirality in all life forms. With only a very few exceptions (in small details) all living cells have similar basic metabolisms and mechanisms for information storage and transfer (Chapter 1). Even body plans of multicellular organisms are variants and developments of the same primary design. It is humbling, but we humans differ by only 1 or 2% of our genes from chimpanzees. We share most of our genes with corn and algae, and our enzymes and body plans are basically the same as those of any fish.

---

[4] This is the present-day figure. For the purpose of this argument, the volume of the early ocean was probably not significantly different.

Table 6.2 Number of species and life forms on Earth

(a) Estimate of the number of extant species of life forms on Earth	
Bacteria	>1 G
of these, cyanobacteria	3600
Algae	33,000
Fungi	900,000
Gymnosperm plants	800
Angiosperm plants	235,000
Total number of animal species	1.4 M
Insects and arthropods	>1 M
Beetles	350,000
Ants and bees	110,000
Fish and amphibians	20,600
Mammals	3700
(b) Number of life forms on Earth with unique metabolisms and cellular-level reproduction mechanisms	1

Life on Earth has been found in exotic places, in the most extreme environments (see Chapter 8). These include dry deserts; the extreme salinity of certain lakes; acidic and alkaline waters and soil; strata 3 km below the surface; clouds 20 km above the Earth; around undersea vents, and elsewhere. Rather disappointingly, for those searching for variety, each and every one of the life forms found in these varied niches is built of the same basic building blocks. This uniformity of life on Earth much simplifies our search for the origin of life, as studies at the genetic molecular level indicate that all life is traceable to just one, currently theoretical, life form. It has been called 'LUCA', the last universal common ancestor. It is currently the holy grail of biologists searching for the origin of life and is thought to belong in the early Archaea.

The singularity of early life raises a rather obvious, but extraordinarily difficult question: *why* is there only one basic life form on Earth?

Some conjectures

Life on Earth is built around the storage of data (in genes) concerning the construction and operation of the life form and the transfer of this information to later generations. Richard Dawkins (in *The selfish gene*, 1976) famously put genes at the very centre of the thrust of evolution. He may have been going a little far in making the librarians more important than the readers, but he made his point—the essentiality of Darwinian natural selection for evolution and the essentiality of evolution for understanding life on Earth.

Genes store information in an extremely simple code. Information handling is based on just four basic building blocks, called nucleotides. These are adenine (A), thymine (T), cytosine (C), and guanine (G). A fifth nucleotide, uracil (U), is used by different forms of RNA, messenger RNA and ribosomal RNA, together with transfer RNA (the DNA–protein intermediary), instead of thymine. Any three of these nucleotides together code for a specific amino acid, and amino acids are built up in certain orders to form specific proteins. It has been suggested that this system is, energy-wise, the least expensive possible. For example, some amino acids have a selective affinity for the base triplets that code for them. According to this approach the system would out-compete other, more energy-intensive life forms and codes which may have evolved in the primordial soup. Moreover, computer simulations suggest that, of the many variants possible, the present code is the one which minimizes transfer errors (Freeland *et al.* 2003).

There are a very few examples of variants to the 'canonical' code, such as the one used for protein synthesis in human mitochondria. Also, there are occasional appearances of another two amino acids, pyrrolysine and selenocysteine, which complete the list of 22 amino acids used by life on Earth. Some fungi and bacteria have unusual codons, such as UGA (usually a Stop codon) for tryptophan and GUG and UUG for Start codons (usually AUG).[5] Even so, despite these rare exceptions, nearly all life is based on the same system of information handling. The question is: why is there no other, perhaps equally simple, code present in other extinct or extant life forms?

Francis Crick, of double helix fame, has suggested the 'Qwerty' explanation. A common experience is the great expense which is often incurred in replacing a known, working technology, with one that may be superior. The cost of the change frequently exceeds the benefit. There are many examples of such 'frozen by accident' technologies. For example, in the United Kingdom many old units are still used for everyday measurements, even though the metric and SI systems are simpler and better defined. The classic case Crick was referring to is the almost universal 'Qwerty' order of typewriter keys. The 'Qwerty' keyboard was originally designed to prevent the clash of levers on mechanical typewriters. With today's electronic keyboards this is irrelevant. But, although there are many known orders of keys which are more convenient and allow for faster typing, it is less costly to stay with the old design. Crick believes that because the original life model worked, any marginally superior code would have been unable to compete with the entrenched system.

Another solution to this problem has been proposed: perhaps other life systems *have* evolved and continue to live on Earth undetected? As indicated in Table 6.2 it is estimated that there are more than a billion species of extant bacteria. Only a small percentage of those that have been observed under the microscope have yielded to laboratory culture, so we know nothing of the metabolism and reproductive mechanisms of the vast majority. The reason for this may be that they are based on other life plans and do not consume agar, the almost ubiquitous substrate used by microbiologists (or any

---

[5] Start and Stop codons indicate from where a sequence along a DNA string, coding for a specific polypeptide, should begin and end.

of the other substrates tried). The possibility of such a 'shadow biosphere' has been discussed by Cleland and Copley (2005) and Davies and Lineweaver (2005) (see also a popular review by Davies, 2007, and the website http://www.astrobio.net/news/article2161/html).

The existence of biological systems different from the one with which we are familiar is of special interest to astrobiologists seeking extraterrestrial life (Chapter 10). Until recently the search criteria have been directed toward environmental conditions and signs of organic molecules of life similar to those on Earth. Science fiction authors have long dismissed this limitation, but established science seems only now to be addressing this possibility, at least in 'thought experiments'.[6] The best summary of such new ideas can be found in the US National Research Council report 'The limits of organic life in planetary systems' (Anonymous 2007).

Hypotheses concerning the origin of life on Earth

There are a number of hypotheses current today about the nature of the first life. As noted above genetic analysis at the molecular level indicates a common origin from the early Archaea, but little more can be said at this time. The first life must have been autotrophic (capable of synthesizing its own organic substances from inorganic compounds, using light or chemical energy). In this it differed from later life forms which were either heterotrophs or autotrophs. Heterotrophs get their energy needs by consuming other, energy-rich organic sources. They include saprophytes, fungi, some bacteria and algae, about 1% of higher plants, and all animals.

Cyanobacteria and most algae and higher plants are autotrophs. They take up and reduce inorganic carbon (usually CO_2) with energy derived from the radiation of sunlight in the process of photosynthesis (see below). The first autotrophs obtained energy from chemical bonds found in certain simple inorganic molecules, such as hydrogen or iron sulphide. These bonds are broken in oxidation, releasing energy which can be used in metabolism, a process called chemosynthesis. It has recently been calculated (Gold 1999; and see Chapter 8) that the greatest mass of all extant life consists of the bacteria which live below the surface of Earth. Consequently, most extant life on Earth is non-photosynthetic and autotrophic. This discovery is very significant in the search for extraterrestrial life (Chapter 10).

Some astrobiologists suggest that instead of liquid water and oxygen produced in photosynthesis being the first criteria in this search (e.g. Kiang et al. 2007) we should rather be looking for the availability of energy sources (Hoehler et al. 2007).

As pointed out by de Duve (2005), a first requirement for the onset of Darwinian selection and evolution is reproduction. So the question of the origin of life comes down to characterizing the smallest, simplest reproducing unit. DNA was quickly ruled out. It is

---

[6] An expression ascribed to, or much used by, Albert Einstein.

simply too large and complicated to have appeared by chance. Moreover, its replication requires the presence of advanced proteins. This led many researchers to prefer a 'RNA first' model. Apart from its 'classic' function in information transfer from DNA (in ribosomes where amino acids are assembled into proteins) RNA has been shown to form ribozymes, groups of relatively small RNA structures which have enzymatic functions. The possibility that ribozymes could replicate RNA lent much credence to the RNA first or 'RNA world' hypothesis.

There is still much criticism of this RNA world theory. Although somewhat simpler than DNA, the RNA nucleotide is quite complex. It contains sugars, phosphates, and nitrogen bases in a very precise construct, which is less stable than that of DNA. The ribose moiety in RNA poses a special problem. It contains a carbonyl group (a carbon atom double bonded to an oxygen atom) which is extremely difficult to synthesize from simple atoms and molecules. According to Steven Benner (see Shapiro 2007) the problem can be solved. Given simple molecules, a boron-rich environment, and lightning, RNA molecules will form. Moreover, Baaske et al. (2007) have shown how nucleotides can be synthesized under conditions simulating hydrothermal ecosystems. Sub-oceanic thermal vents (Chapter 8) have long been considered as candidates for sites where the first life may have originated (Martin and Russell 2007).

In the RNA first hypothesis, as evolution progressed DNA took over the data storage functions of RNA. This was because of its greater stability. Most of the catalytic functions of RNA were later replaced by more specific proteins.

Another hypothesis for the origin of life is 'simple molecules, metabolism first'. According to this concept there was at first considerable interaction (call it metabolism) between relatively simple molecules, which formed groups which slowly became more sophisticated (Shapiro 2007). Simulation modelling of such primitive molecular interactions suggests that this is possible (Shenhav et al. 2007). The first life would be composed of 'ecosystems' of molecules interacting with each other and with the environment (Norris et al. 2007). An additional possibility being considered is that the first life was based on relatively simple catalytic peptides (Fleminger et al. 2005).

The next stage in evolution would be the formation of cells. Oparin and Haldane (see above) were perhaps the first to recognize the essentiality of the separation of the milieu of the metabolizing and replicating units from their environment, by membranes. The first step in this direction is actually not so difficult to conceive, as there are many inorganic systems which form bubbles and cells. The first replicating molecular groups could have found themselves inside such structures. From there, progress towards a functioning cell, surrounded by a selective membrane, would be incremental (Chen 2006). Even so, as of the time of writing, no one has succeeded in synthesizing a replicating cell.

One recent approach to the question of the origin of life is the 'top-down' approach currently being pursued by Craig Venter (famous for his private entrepreneurial participation in the race to sequence the human genome). In this study bacteria are engineered with fewer and fewer genes, with the intention of finding the minimum number which can support a living, reproducing, prokaryotic cell. Gibson et al. (2008), at the Venter

Institute, have reported the first chemical synthesis of the genome of a *Mycoplasma* microbe (a bottom-up approach). If substantiated, this is a breakthrough of dramatic importance. Even so, the path to LUCA is still long.

The origin of photosynthesis

From the first appearance of life about 3.7 Gy BP, and for the next 1.5 Gy (Table 6.1), only autotrophic life reigned on Earth, although signs of anoxygenic photosynthesis have been traced to the late Archean (Xiong and Bauer 2002). As already noted, by sheer mass, this type of life continues to dominate the biosphere (see Chapter 8). From the point of view of fecundity, adaptation to the changing environment, and hence persistence, this early life was, and has been to this very day, extraordinarily successful. From about 2.5 Gy BP, energy-rich reduced molecules, on which these autotrophic prokaryotes rely, became less and less available in the oceans and on the surface of the land. This was the evolutionary driver for the appearance of photosynthesis, which could capture energy from the radiation of the Sun.

In Chapter 1 it was shown how, from the point of view of thermodynamics, photosynthesis changed the biosphere. Formerly it was a closed system in which free energy must eventually become exhausted.[7] Consequently entropy would increase and life cease. With the onset of photosynthesis, the biosphere became an open system, importing energy and enabling a decrease in entropy, which translates into an increase of life.

The evolution of photosynthesis

There are numerous texts on the evolution and mechanism of photosynthesis. Here we relate only very briefly to those aspects that are of importance to the overall theme of how the universe affected and effected life on Earth. Essentially, photosynthesis captures short-wave radiant energy arriving from the Sun. This radiation has been described in Chapter 3, together with the reasons for the apparent inefficiency (here defined as the ratio of light capture to carbon acquisition in plant growth) of the photosynthetic process. The initial photon trapping efficiency of photosynthesis is very high, but when the final carbon productivity efficiency (carbon fixed per unit sea or land area) is calculated it is found to be less than 4% of the total incident solar energy.

---

[7] This is not strictly correct. The biosphere, even without solar radiation, is not a completely closed system. The microflora within the Earth, down to a depth of about 3–5 km, which is the lowest level of the biosphere, receive energy (often molecule bound) from deep within the Earth, and heat from the decay of radioactive isotopes. As for life in the oceans, on land, and in the atmosphere: without solar radiation it would be a closed system. Furthermore the halobacteria, mentioned above as having appeared in the Archaean aeon, used radiation for pumping ions, although they did not have the full electron transport chain of photosynthesis.

It should be remembered that photosynthesis is only possible on Earth because our planet is within the habitable zone of the Solar System (see Chapter 3). This zone is usually defined in relation to the possibility of conditions which support water in the liquid phase. This is certainly a requirement for photosynthesis but, together with this is the requirement for a certain intensity of solar radiation. Obviously, in planets far from the Sun not only would it be too cold for liquid water but there would be insufficient radiation for significant photosynthesis.

Three points are of special interest in the evolution of photosynthesis. The first, noted above, is that anoxygenic photosynthesis was an early event. The second is that the photosynthetic mechanism is extremely complex. It probably came about from the symbiosis of a number of unicellular life forms having metabolic functions quite unrelated to photosynthesis (Margulis and Sagan 1995). This makes its evolution almost as difficult to explain as the early appearance of life itself (Des Marais 2000; Xiong and Bauer 2002). A third consideration of note is that much of the photosynthetic mechanism is highly conserved. The same or very similar elements are present in the most ancient of photosynthesizing algae and in the most modern plants. Take for example ribulose biphosphate carboxylase oxidase (Rubisco). This is the primary enzyme in most extant plants that captures CO_2 from the atmosphere. Only minor changes have been made in Rubisco since its appearance 2.5 Gy BP. It is well adapted for carbon fixation in algae grown in water in equilibrium with air containing high concentrations of CO_2 ($>0.5\%$) or for primitive land plants grown under the same CO_2 level. This was the lowest level of CO_2 in the early atmosphere. Atmospheric CO_2 only fell to $<0.5\%$ in the last 350 million years (see Fig. 7.2). Although the affinity of today's Rubisco for CO_2 is much higher than that of the cyanobacteria it is still much below that required for plants growing in today's atmosphere, which has a CO_2 concentration of $<0.0026\%$. In this context 'today' means the past hundreds of million years, not the period since the Industrial Revolution (Chapter 10). Only in the past ~100 My has a photosynthetic mechanism appeared which has a greater affinity for CO_2 (the C_4 system; see Chapter 8).[8]

In oxygenic photosynthesis, radiant energy from the Sun at wavelengths between 400 and 720 nm is absorbed by chlorophyll and a number of auxiliary pigments. This waveband contains only 48% of the energy available in the solar radiation reaching the surface of Earth below the atmosphere (Chapter 3). Only photons of wavelength shorter than 720 nm have sufficient energy to split the water molecule. This is one of the reasons for the low overall efficiency of photosynthesis discussed in Chapter 3. Hydrogen from water is employed to reduce CO_2 and synthesize triose phosphate. The latter can be used by the plants directly, or is first converted into different disaccharide sugars such as sucrose. These sugars are readily soluble and transportable within the plant.

---

[8] Some red algae do have a CO_2 affinity higher than that of the cyanobacteria, but for unknown reasons this line of algae did not become dominant.

Some of them are then polymerized into insoluble carbohydrates such as starch, which are used for storage.

Radiation at wavelengths shorter than 400 nm is not absorbed by chlorophyll. It can, however, be absorbed by other plant components. In a few cases radiation between 300 and 400 nm (ultraviolet, UV) is absorbed and used for certain special, mainly morphogenetic, functions. Where absorbed, the high energies of these photons, especially at wavelengths below 350 nm, are often sufficient to cause ionization, and are consequently highly destructive and sometimes lethal. About 3% of the energy of the Sun's radiation consists of UV (see Chapter 3). UV is absorbed by DNA. This causes mutations which, as discussed above, are generally damaging but, in very rare instances, beneficial. The latter are an important driver of evolution.

Multicellular life on the surface of Earth is enabled today by the presence of oxygen and its associated molecule ozone, O_3. Together they absorb most of the UV incident on Earth. There is some question as to how the first life was able to become established in shallow seas and on the land surface before the formation of the UV-absorbing (O_2/O_3 containing) atmosphere. It is thought that at that time there were other UV absorbers, such as methane and organic molecules not unlike those composing the haze currently enveloping the satellite Titan (Chapter 4). Moreover, the radiation from the early Sun was about 20% lower than today (Chapter 3). An alternative possibility is that the first life originated in the oceans around deep sea vents. UV radiation does not penetrate to these depths.

The simplest expression of oxygenic photosynthesis is given by the classic equation:

$$6CO_2 + 12H_2O + \text{radiant energy} \rightarrow C_6H_{12}O_6 + 6O_2 + 6H_2O$$

or in words:

$$\text{carbon dioxide} + \text{water} + \text{radiant energy} \rightarrow \text{glucose} + \text{oxygen} + \text{water}.$$

In reality, radiant energy activates the process and is not a reactant. Oxygen is produced as a gas, which is released. Disaccharides or polymers such as starch, rather than the monosaccharide glucose, are the end products. Note that in order for these sugars to be used for energy production they must pass through the respiration pathway. Respiration is the breakdown of these energy rich carbohydrate molecules in a process that produces ATP and nicotinamide adenine dinucleotide phosphate (NADPH). These molecules are the energy 'currency' of plants and animals. Energy is expended in metabolism to synthesize enzymes, cell membranes, cell walls, and all the other components of the organism. A corollary of this is that the evolution of respiration must have pre-dated photosynthesis.

Early respiration was anaerobic. It produced relatively little ATP and NADPH from each energy-rich sugar molecule. Later, in the oxygen-rich atmosphere, respiration could proceed further, producing much more of this energy currency at a greatly increased rate. This was essential for mobility and for large brains, both of which require large amounts of energy (Chapter 7).

Oxygenic photosynthesis first appeared in the cyanobacteria (formerly called the blue-green algae). It should be noted that although the cyanobacteria were the first to produce oxygen they were preceded by at least three other branches of anoxygenic photosynthesizing micro-organisms. They were *Chlorobium* (green sulphur bacteria), *Rhodobacter* (purple bacteria); and *Chloroflexus* (green gliding bacteria). To this should be added the rather exceptional Gram-positive staining heliobacteria. They all used chlorophyll but produced no oxygen. Only the cyanobacteria released oxygen from the splitting of water. The chlorophyll of cyanobacteria is closely related to that of the more advanced eukaryotic algae and higher plants, where it is found in special organelles, the chloroplasts.

The cyanobacteria evolved 2.6 Gy BP. Their activity caused a revolution in the biosphere as a result of the formation of the oxygen-rich 'third atmosphere'. This new period is called the Proterozoic aeon, which followed the Archean. It should be noted that the formation of an oxygen-rich atmosphere was also at least partly dependent on plate tectonic activity, as subduction buried some of the carbon detritus of plants (Chapter 4). Without tectonic subduction, plant detritus would have been broken down by the respiration of soil microflora, which takes up atmospheric oxygen and releases gaseous CO_2. The oxygen ends up in water. As a result, the oxygen concentration of the atmosphere rose slowly (in stages) until it attained the 21% by volume (v/v) prevalent today (Chapter 7).

Consequences of lack of birth control in plants

Population explosion; depletion of natural resources; a toxic effluent poisoning its own milieu and that of other beneficial life forms; empowerment of competing enemies; the cause of detrimental climate change. The list sounds familiar, but it was not posted by Greenpeace to stigmatize humans. It is rather the effect of plants (photosynthesizing, oxygenic life forms) on the biosphere, from the viewpoint not of humans but of plants themselves. In brief, plants were highly successful and prolific. They quickly developed numerous species (Table 6.2a) which spread to all quarters of the oceans and to almost all the land surface of Earth. In doing so they reduced themselves to a state of malnutrition with respect to one of their main food elements, CO_2. They exuded a toxic waste effluent (O_2) which partly poisoned them and from which (with the exception of C_4 plants discussed in Chapter 7) they suffer to this very day. The high-oxygen atmosphere they brought about also poisoned some of their natural resources (especially the anaerobic nitrogen-fixing bacteria) and caused catastrophic climate change (one or more severe ice ages; see Chapter 7). Finally, the high-oxygen atmosphere enabled the evolution and very existence of their natural, competing enemies (insects, parasites, and other animals). You have doubts? Put any leaf under a low-powered microscope and observe the feeding insects. Put plants in a low-oxygen atmosphere (2%, instead of to-day's 21%) or a high-CO_2 atmosphere (0.08%, instead of today's 0.038%) and see how much better they grow! One of the plant world's worst natural enemies, whose life they enable, is *Homo sapiens sapiens*. Think how much better plants would be without us. Witness, for example, the Amazon forest!

Although quite true, the above paragraph has been written somewhat 'tongue-in-cheek'. The intention is to wake up over-ardent 'Greens', with whom the writer is actually sympathetic. When today we attempt to protect and preserve the biosphere in its 'natural state' we must be aware of the human-biased viewpoint. We seek to better the environment—but what does that mean? What exactly is the 'natural' environment? Wars, disease, and famine, leading to a reduction of the human population are quite 'natural' and make good ecological sense. Cockroaches, with their protective carapaces and below-ground abode, would probably enjoy the biosphere in the aftermath of a global nuclear war! Human activity has increased the CO_2 content of the atmosphere by 50% in the last 150 years. Its possible detrimental effect on the climate is debated (see Chapter 9) but it produces a world-wide increase in plant growth. When discussing ecological damage and benefits we must always define to and for whom.

Summary

- First life evolved on Earth 3.7 Gy BP, very soon after the end of the Hadean aeon, during which life, as we know it today, was not possible. There is some evidence for an even earlier appearance.
- The most primitive life forms were extremely complicated. Chance appearance from within the primeval 'soup' in so short a time frame is difficult to understand. One evasive solution is to call upon supernatural creation. Another evasive, but possible solution, is that life reached Earth from elsewhere in the universe (panspermia). Life forms travelling through space must survive dehydration and radiation. This is possible within comets or large meteors, but is unlikely. Moreover, this only transfers the problem of origin to somewhere else.
- Commencement of Darwinian evolution requires reproduction. That DNA molecules were the first life forms to evolve seems unlikely, as they are very complicated in composition and three-dimensional structure and are fragile. RNA is simpler to synthesize than DNA and can act as a catalyst in many protein polymerase reactions. It can also reproduce in RNA-formed bodies—ribozymes. This has lead to the 'RNA first' or 'RNA world' hypothesis.
- Many believe that even the RNA world is too complicated to have evolved by chance. Rather, the formation of simple cooperating molecules and simple polypeptides is postulated as a first possible step.
- Experimental efforts to synthesize organics in the laboratory, simulating the putative early 'soup' and environment, have produced the amino acids used by life, with the correct chirality, and other organic molecules, but nothing approaching a living cell.
- Analysis at the molecular level indicates that all life forms on Earth evolved from the same universal ancestor. They are all based on similar organic molecules, basic structural design at the cellular and whole organ and body level, and method of data storage and replication. There seems to be only one life form on Earth. The reason for this is unknown. The existence of other life forms among the host of un-culturable bacterial species is suspected, but has so far not been demonstrated.
- First life was heterotrophic and used energy sources from molecules produced during the process of formation of the Earth. As these became depleted about 2.5 Gy BP, anoxygenic and then oxygenic photosynthesis appeared, bringing energy into the biosphere which allowed the continuation of surface life and (together with tectonic activity) produced Earth's 'third atmosphere', which is characterized by a high oxygen content.
- The high-oxygen atmosphere was detrimental to plants themselves but enabled the evolution of advanced multicellular life forms, including those extant today.

7

Setting the stage for the evolution of life on a tumultuous planet

When astronauts look back to Earth from space they see a beautiful, unique blue planet, swarming with life. But it is no 'rose garden', either now or in the past. Ever since life first appeared on Earth about 3.7 Gy BP it has had to contend with and adapt to an ever changing and often extremely stressful environment. Some of the more significant of the cosmic and terrestrial factors affecting and modifying the biosphere have already been described. They include: the slowing revolution of the Earth on its axis and its ever changing and complicated circumnavigation of the Sun, which determine and modify day length, weather and climate; the solar radiation incident on Earth, which has gradually increased, by some 20% in the last 3 Gy; the Sun's life-impeding UV radiation, some of which penetrates the atmosphere, and the Moon, which has receded, reducing its originally massive tidal effect (Chapters 3, 4, and 5). In addition, and in part as a result of the above motions, Earth has gone through a series of cold periods, the Ice Ages. Planet Earth itself has not been quiescent, but has challenged the biosphere with movements of tectonic plates (Chapter 4), periodic extreme volcanic activity, upwelling of poisonous gases from the oceans, and changes in the composition of the atmosphere. Finally, in our list of stress factors, Earth has suffered from abuses of extraterrestrial origin, especially the impacts of comets and asteroids. Most recently humans have been affecting the biosphere in ways detrimental to themselves and other species (Chapter 9).

Life did not just adjust to the environment of Earth's biosphere, but often modified it. This is true on both the macro and the micro scales. For example, as described in Chapter 6, plants radically changed the composition of the atmosphere. On a smaller scale, plants also change the environment of the soil around their roots. The effects of bacteria, insects, and animals on the biosphere are legion. Life and the biosphere coevolved (Rothschild and Lister 2003).

Recurring changes in the biosphere have been the driving force for the evolution of new species. This has been by way of rare adaptive mutations, symbiosis, or beneficial combination and activation of randomly accumulated genes (see below). As evolution produced more evolved species, their very complex nature also made them increasingly sensitive to further environmental change. This reduced species' longevity. Whereas Earth has many extant, simple Archaean and Proterozoic life forms, which have flourished for the last 3 Gy, species of the more complex plants and animals rarely persist for more than about a million years.

Despite recurring catastrophes, sometimes on a global scale, life has always survived *somewhere* on Earth. At these times numerous species have failed to adapt and have died out. Sometimes most of the existing life became extinct. Even so, life and evolution never stopped entirely, only to start again.

Earth's 'third atmosphere'

Earth's 'first' and 'second' atmospheres were discussed in Chapter 5. The evolution of photosynthesis and its overall effect on the composition of the atmosphere were described in Chapter 6. In brief, photosynthesis caused a large rise in oxygen and a drop in CO_2, producing Earth's third atmosphere.

From the onset of oxygenic photosynthesis towards the end of the Archean aeon (about 2.5 Gy BP see Table 8.1 for the dating and characteristics of the different geological periods) the oxygen content of the atmosphere began to increase (Fig. 7.1a). After the appearance of the first cyanobacteria it rose from just a trace to ~1% v/v,[1] then to ~5% during the Proterozoic aeon. After the Proterozoic the oxygen level rose rapidly. From the beginning of the Phanerozoic aeon (which began with the Cambrian 'explosion' of 542 My BP) the oxygen level 'soon' reached 21%. With some highly significant fluctuations, it remained at this level to the end of the Quaternary period (i.e. today).

The rise in the oxygen content of the atmosphere from the 1–2% of the first atmosphere to the 21% of the Phanerozoic was caused by the combined effect of photosynthesis (Chapter 6) and sequestration of organic material derived from photosynthesis and its metabolic products. This sequestration was often the result of subduction or burial caused by the movement of tectonic plates (see Chapter 4).[2] Two other processes can sequester carbon and bring about an increase in the oxygen level. The first is the formation of precipitates of organic origin, usually in the form of calcium carbonates which do not dissolve but sink to the bottom of the seas. The second is the activity of diatoms. Diatoms are unicellular or colonial yellow-brown algae. They appeared some 500 My BP and spread rapidly, diversifying into thousands of species dispersed throughout the oceans. They have extremely stable silicon-containing cell

---

[1] The gas contents of the atmosphere are given in the commonly used units of % v/v (per cent by volume). The SI unit for pressure is the pascal (Pa): 1 Pa = 1 N m^{-2}. A content of 20% O_2 v/v is approximately equivalent to 20 kPa; 350 parts per million (ppm) v/v (or 0.035% v/v) CO_2 is equivalent to 35 Pa.

[2] When this happens the organic matter may be converted into coal or oil by heat and pressure. These 'fossil fuels' usually remain within the Earth until burnt by humans. Burning fixes oxygen and returns CO_2 to the atmosphere.

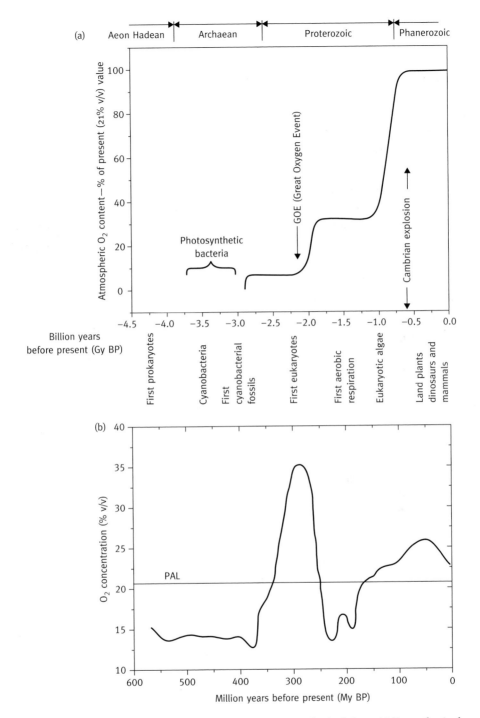

Fig. 7.1 The increase in Earth's atmospheric oxygen over geological time. (a) From the Archaean to the Quaternary; (b) over the Phanerozoic (PAL, present atmospheric level). Some of the details shown in these curves are being debated, but the overall trends are generally accepted. Curves based on data collated (and smoothed) from many different sources.

walls. When they die they sink to the ocean floor, taking with them much of their cellular carbon.

Photosynthesis releases oxygen by splitting water molecules. As long as the total amount of organic tissue produced in plant metabolism increases (as for example in the bodies of living algae and higher plants, in sequestered organic material, or in ocean precipitates) the oxygen content of the atmosphere will rise, while CO_2 decreases.

In a balanced ecosystem, such as today's remaining areas of virgin Amazon forest, the total plant mass remains constant as the newly produced primary plant production (leaves, fruit, etc.) is consumed. It falls to the ground as detritus, where it is decomposed by the soil microflora. In so doing, the respiration of the microflora returns CO_2 to the atmosphere, while taking up oxygen. Consequently, and contrary to a common misconception, a constant stable plant ecosystem does not affect the oxygen and CO_2 content of the atmosphere. This is true unless there are large-scale natural fires or the vegetation is cut and burned by human activity.

As shown in Fig. 7.1(a), by the beginning of the Phanerozoic the oxygen content of the atmosphere had risen to more or less its present level of 21% v/v. However, as shown in Fig. 7.1(b), it has varied over these last 500 My in the range of 15–35% (Berner 1999). When oxygen rose to its highest levels, in the Carboniferous period, it caused worldwide vegetation fires (see below).

Oxygen, UV radiation, and early ice ages

Earth's 'second atmosphere' (Chapter 5) contained high levels of CO_2 and methane (CH_4). Methane is produced mainly by anaerobic bacteria which prefer temperatures of ~40°C. It has a much higher absorption of infrared radiation than CO_2 and consequently was (and is) a very important greenhouse gas, despite there usually being far less of it. In Earth's second atmosphere their combined effect was to raise the surface temperature of Earth to values compatible with life, supporting liquid water; this was at a time when solar radiation was some 30% less intense than it is today.

As the oxygen level rose the methane was oxidized. This was mainly the action of the hydroxyl (OH) and superoxide (O_2^-) radicals, produced by the effect of solar radiation in the oxygen-rich atmosphere. As a result, and at least three times in the Proterozoic, between 2.3 and 2.2 Gy BP and later between 750 and 600 My BP, the temperature of Earth dropped dramatically. The fall was so severe as to bring about extreme ice ages (Kopp et al. 2005). These have been called periods of 'snowball Earth'.[3] To what extent the ice approached the equator is still being debated (for a popular account see Walker, 2003). Certainly not all Earth's water froze, as life was

[3] Other factors have been implicated in these severe events, foremost among them being the position of the land masses which, at that time, were close to the equator. This would result in an increase of Earth's albedo at higher latitudes and hence cooling.

never entirely extinguished. A model developed by Goldblatt *et al.* (2006) from physical considerations (reviewed by Kasting, 2006) predicts a cycling between ice ages and interim warm periods in the Proterozoic, as also indicated by the geological evidence. There were many later periods of ice ages, of different degrees of severity and duration, with quite different causes (McDougall 2004). They are discussed below.

The importance of UV attenuation to life was discussed in Chapter 3. By the time of the Phanerozoic most solar UV was being absorbed by the oxygen and ozone in the atmosphere. The amount of UV incident on Earth's surface before this period is much more uncertain (Cockell 2002).

Some biological effects of the oxygen revolution

The negative effect of the high-oxygen atmosphere on plants was pointed out in Chapter 6. The same high oxygen level also allowed the emergence of advanced complex life, as explained below. The many effects, problems, and benefits to life, resulting from the high-oxygen atmosphere have been reviewed in two recent books, Lane (2003) and Beerling (2007) and in a paper by Lenton (2003).

Oxygen and nitrogen availability

One of the first negative effects of the high-oxygen atmosphere was the inhibition of nitrogen fixation by anaerobic bacteria. Fixation is carried out by nitrogenase enzymes which are inhibited by even low levels of oxygen. Shortage of available nitrogen is one of the main limiting factors in plant ecosystems. A shortage of nitrogen-containing foods, such as proteins, is a major limitation to the growth of many species of animals: this includes humans in many regions of the world, especially sub-Saharan Africa.

The cyanobacteria constitute only ~1% of the oceanic algae. When they evolved their most significant feature was oxygenic photosynthesis. The mechanism for this was later installed, by invasive symbiosis, into other life forms which evolved into photosynthesizing plants (Chapter 6). Their second important characteristic was nitrogen fixation, a facility present in many species of cyanobacteria. This they accomplish by a number of strategies, all of which separate the nitrogenase enzymes from oxygen. Some cyanobacteria produce special cells which cannot photosynthesize, called 'heterocysts', in which the oxygen concentration is low. Others carry out nitrogen fixation only in the dark, when there is no photosynthesis and, again, cellular oxygen is low (Kasting and Siefert 2002). As discussed above, the higher leguminous plants (Leguminosae) also separate the nitrogenase of their nitrogen-fixing symbiotic bacteria from oxygen. In this case the bacteria are sequestered in oxygen-excluding root nodules. Apart from the cyanobacteria and Leguminosae with their nitrogen-fixing symbiotic mechanisms only a few other plants can obtain atmospheric nitrogen. For example, the roots of alder trees (family Betalaceae) also have a symbiotic relationship with anaerobic nitrogen-fixing bacteria in flooded soil where the oxygen level is low. Other plants must rely on nitrate and ammonium ions, where available.

Soils contain nitrogen-fixing bacteria which, when they die, supply nitrogen in forms which can be taken up by plants. A few of these bacterial species can live in a high-oxygen atmosphere, but most are anaerobes. Some insects and animals, such as ruminating cows, have nitrogen-fixing bacteria in the anoxic environment of their guts, but the quantitative importance of this nitrogen source is unclear.

Oxygen and respiration

Respiration is the mechanism by which cells mobilize the energy stored in molecules produced (basically) by photosynthesis or captured heterotrophically. The most important energy 'currency' produced in respiration is ATP. All metabolic activity and the pumping of various molecules against chemical gradients, through membranes, require energy derived from ATP.

Anaerobic respiration evolved before the onset of the high-oxygen atmosphere. Basically it can be described as glycolysis, the breakdown of glucose to pyruvic acid, which releases energy:

$$glucose \rightarrow pyruvic\ acid$$
$$C_6H_{12}O_6 \rightarrow C_3H_4O_3 \quad (releasing\ 2\ ATP).$$

This is the final reaction. Pyruvic acid is the first product. In fermentation, a variant of anaerobic respiration such as occurs in yeast (a facultative[4] aerobe), ethanol or lactic acid are produced.

One of the most important effects of the high-oxygen atmosphere ($>2–3\%$ O_2) was to enable the evolution of aerobic respiration. In aerobic respiration the pyruvic acid, lactic acid, or ethanol (produced by glycolysis) is further oxidized. The final products are CO_2, water, and much more ATP:

$$C_6H_{12}O_6 + 6O_2 \rightarrow 6CO_2 + 6H_2O \quad (releasing\ 38\ ATP),$$

The exact quantity of energy produced in aerobic respiration is not certain, especially in eukaryotic cells where the post-glycolysis reactions take place in the mitochondria. In these cells a small part of the energy is expended in moving pyruvic acid and other respiratory intermediaries through the mitochondrial membrane. In prokaryotes all the metabolic steps of respiration take place in the cytoplasm. Even with these energy losses, aerobic respiration is about 19 times more efficient in releasing energy from sugars than anaerobic respiration. This is the main factor which enabled the evolution of insects and large animals. Unlike algae and higher plants, mobile advanced life forms require the availability of relatively large amounts of energy. As noted above, the first annelid worms may have appeared in the late Archaean. In comparison with insects,

---

[4] Capable of switching between aerobic and anaerobic respiration in response to the level of available oxygen.

for example, they do not require much energy, but certainly require far more than most prokaryotes. Consequently, the annelids could benefit from aerobic respiration which evolved at about the same time.

Oxygen toxicity to plants and animals

The scale used in Fig. 7.1 suggests that the amount of atmospheric oxygen in the Phanerozoic was constant at about 20% v/v. In fact, as discussed previously, various proxies for oxygen suggest that there were periods, especially in the Carboniferous (360–286 My BP), when it rose to as high as 35%. One such indication, mentioned above, was the world-wide prevalence of fires which led to the formation of coal, increased the CO_2 content and reduced the oxygen content of the atmosphere (Watson *et al.* 1978; Scott and Glasspool 2006). (See Lane, 2003, for a popular but very detailed account of the effects of a high level of atmospheric oxygen on life.) Dudley (1998) discusses the evolutionary responses to both a high oxygen level and a putative denser atmosphere in the late Palaeozoic. He suggests that this was the reason for the appearance of giant early flying insects, arthropods, and tetrapod locomotion. A late Permian hypoxia eliminated many of these giant species.

One of the conundrums of palaeobiology is associated with the catastrophic event, further described below, which delineated the K/T boundary 65 My BP. At this time many of Earth's living species became extinct. One outstanding puzzle is that part of the flora and fauna of the time began to deteriorate, starting about 70 My BP, i.e. some 5 My before the main extinction event. At that time the oxygen content of the atmosphere is thought to have been as high as 28–35% v/v, while the CO_2 content was as low as 0.03%. Rachmilevitch *et al.* (1999) found that leaf stomata respond to this atmospheric composition by opening wide. The result is a high conductance for gases and a reduction in the ability of the vegetation to regulate water loss. This would have led to water stress and reduced plant growth in many regions with an arid climate. Gale *et al.* (2001) reported that the same high oxygen and low CO_2 content could have further reduced plant growth by partially inhibiting net photosynthesis. If this reduction in plant growth did occur it would have affected the growth of all plant-dependent heterotrophic insects and animals.

As discussed above, the current level of oxygen (21% v/v, or 21 kPa) is to some extent toxic to most plants and to all animals. It often produces free radicals[5] of oxygen which, among many other negative effects, are suspected of reducing longevity (Lane 2002). One toxic effect of oxygen on plant photosynthesis occurs under conditions of strong sunlight and cold temperatures. This photo-oxidation can permanently damage the photosynthetic mechanism. Another negative effect of oxygen is that it directly

---

[5] Free radicals are atomic or molecular species with unpaired electrons. They are highly reactive, often with damaging results to living cells.

competes with CO_2 for the active binding site on Rubisco, the primary enzyme which captures atmospheric CO_2.

This enzyme was originally called ribulose biphosphate carboxylase (RuBC). With the discovery of its affinity for oxygen (especially significant under the present high atmospheric levels), it is now called RuBC-oxidase, or Rubisco. The affinity of Rubisco for CO_2 is very much higher than for oxygen, but this is balanced by the vastly greater atmospheric partial pressure of oxygen (21 kPa versus the 38 Pa of today's CO_2—or 21% v/v versus 0.038% v/v). The combination of high oxygen and low CO_2 in the present atmosphere compared with that prevalent at the time photosynthesis first appeared reduces net photosynthesis by some 30% below its potential. Note that photosynthesis shows at least an initial linear response to high CO_2, while growth is exponentially related to photosynthesis. This means that plant growth may be inhibited by far more than 30% under conditions in which the rate of photosynthesis is a limiting factor. Only 'recently' has a new carbon-fixing mechanism evolved which solves this double problem. In this case 'recently' means at the most 150 My BP and possibly only 15 My BP. This mechanism is called C_4 photosynthesis (see Chapter 8).

Carbon dioxide in Earth's third atmosphere

The level of CO_2 in the atmosphere of the Archean (3.8 to 2.5 Gy BP) is thought to have been >4% v/v. Estimates of its level in the Proterozoic (2.5 G to 545 My BP put it at about 3.6% (Rye *et al*. 1995). There are rather more reliable proxies for estimating CO_2 in the Phanerozoic (from 545 My BP to the present). The integrated values for the Phanerozoic, obtained from a number of authors, especially Berner (1997) and Beerling and Berner (2005), are given in Fig. 7.2. Note that in the early Phanerozoic the CO_2 level was still of the order of 0.6%. This was more than 20 times the pre-industrial level of 0.025% v/v (i.e. before the mid 19th century). It has been more or less constant (with only occasional digressions) over the last 350 My.

The effect of oxygen on the primary CO_2-fixing enzyme, Rubisco, was noted above. Rubisco was well adapted to the levels of oxygen and CO_2 present before the Cambrian (<2% O_2 and >3% CO_2) but not to today's atmosphere. The Michaelis–Menten coefficient, K_m,[6] of Rubisco for CO_2 is about 200–400 μmol. This is much higher than the 10 μmol equilibrium level in water at a pH of 7.0 and temperature of 25°C (free, or in plant leaves) in contact with air containing CO_2 at a level of 0.03%.

Rubisco evolved in early photosynthesis and its K_m for CO_2 has changed only very moderately since. This lack of evolution is probably related to a lack of evolutionary pressure.

---

[6] The Michaelis–Menten constant, K_m, is defined as the substrate concentration at half maximum velocity of an enzyme-facilitated reaction. Consequently, the lower the value of K_m the higher the affinity of the enzyme for the substrate. The K_m value obtained for Rubisco varies according to the method used for analysis. It is somewhat lower when coming from plants of more recent lineage, but the difference is not great.

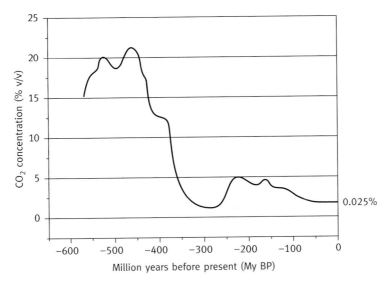

Fig. 7.2 Carbon dioxide levels in Earth's atmosphere over the Phanerozoic. See text for sources of data.

Apart from a period between 350 and 260 My BP and the recent Tertiary and Quaternary periods, atmospheric CO_2 was >0.1% (Fig. 7.2). At this level Rubisco functions well. The role of CO_2 in plant evolution has been reviewed by Chaloner (2003).

The present level of CO_2 in the atmosphere (~0.038% v/v, or 38 Pa) has little effect on animals. The level of CO_2 has to rise to about 5% before it produces any physiological response from mammals such as humans. Although the data depicted in Figs 7.1 and 7.2 show little variation in the late Phanerozoic, there were sometimes very significant changes. The levels of oxygen and CO_2 changed with the coming and going of ice ages and tectonic activity. The dramatic anthropogenic increase in CO_2 over the last 150 years, and its effects on the biosphere, are discussed in Chapter 9.

Composition of Earth's atmosphere in the late Quaternary period

The composition of the atmosphere prior to the Industrial Revolution is given in Table 7.1. Two of the most biologically important components of Earth's third atmosphere, oxygen and CO_2, were discussed above. The air of course contains other gases of significance to life, the most important of these being water vapour and nitrogen.

Water in Earth's atmosphere

The atmospheric component which shows the greatest variability is water vapour (Table 7.1). It may comprise as little as 0.1% over dry deserts, rising to 3% (3 kPa) in

Table 7.1 The main components of Earth's third atmosphere prior to the Industrial Revolution

Component	Quantity (% v/v)
Nitrogen	78
Oxygen	21
Water vapour	0.1–3.0
Argon	0.93
Carbon dioxide	0.025
Neon	0.0018
Helium	0.0052

As noted above, values are given in the commonly used units of % v/v, but the SI unit for pressure is the pascal (Pa). At sea level the equivalent value for 21% v/v O_2, for example, would be a partial pressure of 21 kPa and for 0.025% CO_2, 25 Pa. Water vapour is the most variable component, and the 3% v/v value shown (3 kPa) is for conditions close to saturation, as occurs in foggy or rainy weather. Its presence affects the percentage content of the other gases but not the independent partial pressures. The CO_2 figure is for the year 1850. It is currently (in 2009) 0.038%, and rising rapidly. Argon, neon, and helium have no known biological activity.

saturated air (such as during fog or rainfall). Water vapour content is an extremely important factor in the water use, growth, and survival of land plants. In this it is of importance second only to solar radiation.

Clouds are also of great importance to life, but their water content is in liquid drops. As such they are not actually part of the atmospheric *gas* content, but they strongly affect the Earth's energy balance. Solar radiation incident on Earth is partly reflected back to space and partly absorbed by clouds. Moreover, together with aerosols and atmospheric dust, they also contribute to the greenhouse effect by absorbing part of the long-wave radiation emitted from the surface of the Earth (Figure 3.4 and Chapter 3).

Nitrogen in Earth's atmosphere

As shown in Table 7.1 the most abundant gaseous component of the atmosphere is nitrogen (N_2). Gaseous nitrogen is relatively inert and has little if any effect on life. However, life is entirely dependent on nitrogen-containing compounds. It is a main component of primary molecules such as nucleic acids, amino acids, and proteins. Plants and animals cannot use gaseous nitrogen directly (Mancinelli 2003). As discussed above, a major effect of the high-oxygen atmosphere was to severely limit the availability of nitrogen to living organisms. This was a result of the inhibition of nitrogen fixation by anaerobic nitrifying bacteria. Most organisms can only use nitrogen bound in inorganic or organic molecules. Before plants can absorb nitrogen it must first be reduced to NH_3 or NH_4^-. On the early Earth some nitrogen derivatives were produced in the atmosphere by lightning (as is the case today) or arrived in organic molecules within comets impacting Earth. These were relatively minor sources.

Heterotrophic life forms such as insects, animals, and a few parasitic or insectivorous plants obtain fixed nitrogen from consuming plants or other insects and animals. Land plants growing in tropical regions frequently suffer from a shortage of nitrogen. This is because the forms of inorganic nitrogen in the soil which can be taken up by roots, are very soluble. The heavy rains of the tropics frequently flush the nitrogen to soil levels below the plant roots.

Major stress factors which affected the evolution of life

Ice ages

Ice ages are periods during which the average surface temperature of Earth drops and snow and ice advance from the poles towards the equator. The severe pre-Phanerozoic ice ages were discussed above. They were caused mainly, but not entirely, by the early high-oxygen atmosphere which reduced the greenhouse effect of the atmosphere by oxidizing its then high methane content. One of the severest of the ice ages receded only a few million years prior to the Cambrian period (~550 My BP). This warming no doubt contributed to the huge increase in life forms at that time. Apart from one major ice age in the Permian, the early Cambrian to the late Quaternary period was relatively free of ice ages. This was an important enabling factor in the world-wide evolution of life. There were even periods when conditions suitable for plant and animal growth approached the poles, as evidenced today in the oil and coal deposits found at far northern latitudes (see below for further discussion). However, another factor in these anomalies was that land masses were moving around as a result of the motion of tectonic plates.

In the Quaternary period (1.6–0.1 My BP) the ice ages returned, but in relatively short cycles, separated by interglacial warming periods. Reliable data on these occurrences over the last half million years have been obtained from ice cores from deep drilling shafts near the poles. These ice samples contain gases, the isotopic concentration of which is a proxy for the temperature under which they were entombed. Much of our knowledge of the temperatures of the last 430,000 y of the Quaternary comes from the ice shaft drilled by the Russians at their Vostok Antarctic station (Fig. 7.3).

As shown in Fig. 7.3 there have been a series of temperature cycles each lasting approximately 100,000 y. It should be noted that the temperature range between the lows of the glacial to the highs of the interglacial periods is less than 10°C. Also of interest is the rapidity of the exit from the periods of temperature lows. Earth is only now emerging from the last ice age which came to an end some 11,000 y BP. There can be little doubt that the cultural development of humans in the last 10,000 y owes much to this warming, as before this period most of northern Europe was covered in ice. Today we are concerned with global warming. It should, however, be borne in

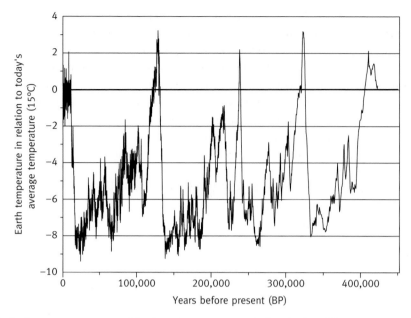

Fig. 7.3 Surface temperatures of the late Quaternary period, calculated from gas isotope ratios in ice cores extracted from drill holes at the Russian Vostok Antarctic Research Station. The line through 0 on the y-axis represents today's average Earth temperature (15 C).

mind, that the observed and worrying recession of the glaciers and ice sheets is part of this ongoing emergence from the last ice age. Whether this global warming is unexpectedly rapid and is being accelerated by human activity is discussed in Chapter 9.

The reason or reasons for the cycling in Earth temperatures as found in the Quaternary period is uncertain. There are many theories such as 'Milankovitch cycling'. Milutin Milankovitch was an early 20th-century Serbian mathematician who studied the combined effect of the three perturbations in Earth's circumnavigation of the Sun (precession, obliquity, and eccentricity; see Chapter 5) on the solar radiation incident during different seasons on different regions of Earth. Milankovitch calculated that there would be a 100,000 y cycle during which the tilt of the Earth would lead to a greater or lesser solar radiation flux. This should be especially noticeable at northern latitudes and could be responsible for the advance or retreat of northern glaciers. Originally there was much scepticism as to the strength of this effect. However, the Vostok data (Fig. 7.3) were found to be in reasonable agreement with his predictions—although this is at best a correlation, not a proof.

Ice ages tend to be self-reinforcing. This is because the growing snow and ice cover increases the Earth's albedo, thus reducing the net short-wave radiation absorbed by the Earth. An important aspect of ice ages is to understand the factors which counter this effect. A possible feed-back temperature control effected by tectonic plate activity was discussed in Chapter 4. Another factor which would tend to counter ice ages is the greenhouse effect of atmospheric CO_2. As temperatures drop so does the rate of world-wide photosynthesis. This leads to an increase in CO_2 and the greenhouse effect.

There are other possible explanations for the cycling Earth temperature. One of these takes us back to the importance of the placement of the Solar System within the Milky Way (Chapters 3 and 4). Shaviv (2003) calculated that the cycling passage of the Sun through the Orion arm of our galaxy modifies the cosmic radiation in the Solar System. As discussed above, this can modify cloud formation on Earth, increasing or decreasing the Earth's albedo. Shaviv also argued that the cosmic radiation flux varies with the rate of star formation in the Galaxy, as calculated by other authors. A paucity of star formation, in the period 2–1 Gy BP, correlates with an absence of ice ages at that time. Shaviv's surprising hypothesis has recently been supported by the work of Gies and Helsel (2005).

Comets, asteroids, and extinctions

Comets and asteroids sometimes leave their usual orbits and enter the inner Solar System, approaching Earth (Chapter 4). As in the K/T event, when Earth was struck by one or more comets or asteroids (see below), the danger to life could be very great. Unlike the situation on Mars or the Moon, where there is little or no weather, most impact sites on Earth are soon covered over as a result of erosion. One such asteroid impact site which, thanks to desert conditions, has survived almost intact, can be seen in what is now Arizona (Fig. 7.4). Other large or larger craters, but not so well preserved, have been found at other sites on the planet's surface. For example, a crater in Chesapeake Bay, off the eastern coast of the USA, which has been dated to 35 My BP.

The crater in Arizona shown in Fig. 7.4 was formed from the impact of a nickel–iron asteroid 50,000 y BP. The asteroid is estimated to have been 50 m in diameter. Before reaching Earth it weighed ~300,000 tonnes, but half vaporized as it entered the Earth's atmosphere. The impact produced an explosion variously estimated to have been equivalent to 2.5 to 15 megatonnes of TNT (>150 times that of the atomic bomb dropped on Hiroshima). It scattered debris over 260 km^2. A larger asteroid would have had world-wide effects on the biosphere, as did the K/T event. Even an asteroid the size of the Arizona meteorite would be capable of destroying entire metropolitan areas should it strike at their centre, causing millions of casualties.

We tend to think of such calamities as the K/T or Arizona events as belonging to the far distant past and not of present concern. However, such calamities could occur today. A warning of about 2 years may be all that is possible, even given astronomers' present observational abilities.

Much nearer to our time such an event did happen, in June 1908, in Tunguska, a fortunately sparsely inhabited region of Siberia. Eighty million trees were burnt or

Fig. 7.4 Meteorite crater in Arizona. Note the size of the approach road. Courtesy of Wikipedia.

knocked over in an area of 2150 km² (Fig. 7.5). A short time after the impact its noise and the resulting fluctuations in air pressure were recorded as far away as London. No trace of an impactor has ever been found. It is thought to have been caused by a small comet or meteorite that vaporized while still at an altitude of 5–10 km. The resulting explosion is estimated to have been the equivalent of ~5 megatonnes of TNT. Dust rose to the stratosphere and reduced its transparency for months. Again, as for the Arizona meteorite, had this body struck a large urban centre the effect would have been horrendous.

If the Tunguska event also seems a long time ago (especially to the very young!) it should be remembered that we are constantly being threatened by such impacts. A bolide as large as the one which hit Tunguska is calculated to strike Earth *on average,* once in 300 y. Recently an asteroid, given the name Apophis (after an ancient Egyptian demon), was calculated to be in an orbit which would impact Earth in either 2029 or 2036. Fortunately, a revised estimate indicates that it is very unlikely to strike Earth this time around. If it does strike Earth its explosive equivalent will be 100 times greater than that of the Tunguska event. There are at least 10 asteroids with a width of around 5 km in Earth-crossing orbits.

Fig. 7.5 Aftermath of the 1908 Tunguska event. Courtesy of Wikipedia.

Extinctions and our position in the Milky Way

Leitch and Vasisht (1998) proposed that periodic mass extinctions could be related to the Sun's passage through the spiral arms of the Galaxy. They postulated that as the Sun passes through dense regions of stars the probability of near encounters with supernovae, and perturbations resulting from the gravitational attraction of other stars, greatly increases. The effect would be periodic (a ~26 My). They correlated this with the palaeontological extinction record which appears to show such periodicity (Raup and Sepkoski 1984). While the patterns found in these fossil data records were previously challenged, they have recently received further support (Rohde and Muller 2005).

Supernovae

The prevalence and possible effects of supernovae (some good) were discussed in Chapters 2 and 3. There have been some 100,000 supernova events in the Milky Way since its formation. It is sufficient to note here that had there been such an event within range of Earth all life would have been incinerated. This means that Earth has been spared in the last 3.7 Gy. Tomorrow may be different. In early 2008 an exploding supernova was observed from many places in the world over many hours. The starlight flash which reached Earth was the brightest ever recorded. It might have been seen with the My naked eye, although there were no such reports. It came from another galaxy at the merciful distance of 7.5 G ly—halfway to the end of the observable universe. Had it been 4 G ly closer life might never have arisen!

Volcanoes

The cause of volcanism and its relation to the activity of Earth's tectonic plates were discussed in Chapter 4. Volcanoes have featured in many cases of local and even world-wide events which have (usually but not always) negatively affected life on Earth. They were the source of Earth's original atmosphere and continue today to add CO_2, sulphur dioxide (SO_2), and other gasses. Their eruptions affect the local biosphere. From time to time they have produced world-wide life-stressing conditions (see the discussion of the Deccan Traps in the section 'Mass extinctions, in the fossil record' below). Another such massive volcanic event took place in the mid-Cretaceous period, about 125 My BP. It produced huge lava flows that are now under the western Pacific Ocean. As a result the sea level was raised by 250 m and Earth's average temperature increased by as much as 10°C (Caldeira and Rampino 1991). Massive volcanic eruptions affecting the entire world are suspected of causing some of the Phanerozoic extinction events (Bindeman 2006). Very small eruptions, in the form of undersea vents, support unique ecosystems (Chapter 8) and may have been the site of the first life on Earth. Other than on Earth, active volcanoes are quite rare in the Solar System, although there are many signs of past eruptions, on Mars for example. Only Io, the innermost of Jupiter's four Galilean satellites, has more extant volcanic activity than Earth.

Toxic out-gassing from the oceans

Last, but not least of the possible contributors to world-wide extinctions is the occasional sudden and large-scale exudation of toxic hydrogen sulphide from the oceans. This occurs when there is an increase in the temperature of Earth and anoxia in the oceans. It is an event which is seen today, on a small scale, in the sea off Namibia on the west coast of southern Africa. Hydrogen sulphide is toxic to both plants and animals. It is suspected of having produced global disasters in the past (Kump *et al.* 2005), and could do so again if the global temperature rises sufficiently.

Mass extinctions in the fossil record: local and planet-wide catastrophes

Palaeontologists derive the history of life on Earth from the study of fossils. Most of the earliest fossils that have been found represent the multicellular organisms which developed after the Cambrian 'explosion' some 550 My BP. The results from the study of these and other fossils, of both ocean and land flora and fauna, indicate that life went through a series of extreme crises. They were frequently of a very short time span ('short' in this case, meaning 1–2 My). At such times large percentages of species and families disappeared from the fossil record or were greatly reduced. Many of these events delineate the accepted geological divisions (Fig. 7.6).

As shown in Fig. 7.6, five major extinctions are manifested in the fossil record during the 500 My of the Phanerozoic. The greatest of these took place 252 My BP and brought an end to the Permian period. In a 'short' period of about 1 My some 70% of all land species and 85% of marine species disappeared from the fossil record. The same overall picture, but of somewhat reduced severity, took place four more times, and there were at least 15–17 relatively minor extinction events. 'Minor', but, as in war, for the non-survivor the damage is 100%!

The jury is still out as to the cause(s) of the end-Permian catastrophe. Almost all the factors mentioned above have been implicated: terrestrial, such as plate motion causing extreme volcanism and impacts from extraterrestrial bodies, such as large meteorites or comets. Evidence for the latter are a number of appropriately dated crater remains found in southern China and in Shark Bay, Australia. Two recent books have been devoted to this traumatic event (Benton 2003; Erwin 2006).

The mass extinction which has received most popular scrutiny is the one that took place 65 My BP and delineated the Cretaceous from the Tertiary period. It is called the K/T boundary event ('K' to differentiate it from the 'C' of the Carboniferous, by using the German word for Cretaceous *Kreidezeit*). The special attention given to this

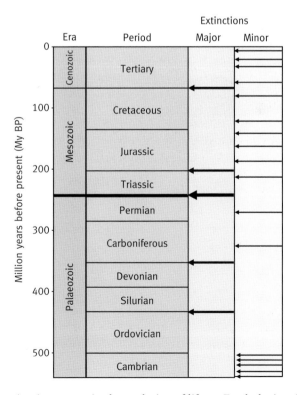

Fig. 7.6 The main extinction events in the evolution of life on Earth during the Phanerozoic, as revealed by the fossil record. Redrawn from Courtillot (1999).

is probably a result of two factors. The first is that the K/T event brought about the end of the period when dinosaurs were the largest animals on Earth (Fig. 7.7). Dinosaurs have a special place in human culture, featuring in many children's (and adult) horror stories and movies. Of greater evolutionary significance is that some 70% of the other species perished, including all animals heavier than about 20 kg. Their demise enabled the expansion of the mammals and eventually the appearance of humanoids and *Homo sapiens sapiens*. The expansion of the mammals was not immediate, as first thought but began some 10 My after the K/T event. Among the dinosaur survivors were at least some avian-like species from which modern birds evolved.

The second reason for the popular interest in the K/T event is that, for a decade, the media delighted in a debate raging in science circles as to its cause. Demonstrating that scientists too are human, the argument often descended to very personal levels. At the crux of the argument was whether the extinction was a result of terrestrial or extraterrestrial factors. Prior to 1970, palaeontologists had ascribed the K/T catastrophe to a number of factors of terrestrial origin, especially volcanism. Underlying India today

Fig. 7.7 Artist's impression of a dinosaur; one of a family which dominated life on land before the K/T event. Courtesy of Chicago University Press.

is a vast lava field called the Deccan Traps. This was formed by a gargantuan volcanic event which was dated to 65 My BP. This volcanic activity must have caused world-wide changes in the atmosphere, such as an increase in CO_2, dust, and SO_2. It is unclear whether the extra greenhouse gases in the atmosphere (such as CO_2), would have increased world temperatures or whether the temperature dropped as a result of an increase in the albedo of the atmosphere caused by sulphur aerosols. Increased albedo would have been caused by the short-wave reflectivity of the dust, aerosols, and cloud cover. Whichever it was, plant life and hence dependent animal life, would have been severely affected by the reduced sunlight, by changes in temperature, and by the precipitation of acidic rain.

In the late 1970s the geologist Walter Alvarez and his father Luis (a Nobel prize-winning physicist) reported that rocks dated from the end of the Mesozoic, in the K/T period, contained a very high concentration of the element iridium, which is present in meteorites but is very rare on Earth. They and their colleagues found iridium anomalies in strata spanning the K/T divide at many sites around the world, at a concentration some 30 times higher than could be expected if the source was terrestrial. Moreover, the same thin dividing strata often contained fractured quartz beads, which are formed under the high pressures and temperatures associated with impacts. From this and other evidence, they deduced that at that time Earth had been struck by a bolide of 5–15 km in diameter. An impact site of the right dimensions and date has been identified in what is now the Yucatan Peninsula of Mexico. It caused a 170 km wide crater, much eroded but still visible in airborne radar pictures. It has been given the name of Chicxulub. There can be no doubt that such an impact, which has been calculated to have had the force of millions of atomic bombs, would have strongly affected global weather and life. The effect could have lasted for a few years, reducing plant growth and dependent life.

The Alvarez hypothesis was at first greeted with derision. Part of this was no doubt resentment at the invasion of palaeontology by mere physicists. However, to the credit of international science, the findings of the Alvarez team are now accepted. Even so, and as described above in relation to variations in atmospheric O_2 during the pre-K/T period, there were certainly other factors which brought about a decline in the biota *prior* to the Yucatan impact event. In that period there were times of high atmospheric oxygen which would have much reduced plant growth and dependent animal life (see above and Gale *et al.*, 2001). There were also times when the oxygen level fell to below 20%. This would have caused severe respiration problems for some of the larger dinosaurs. Whatever happened, there are clear signs in the palaeontological record of a decline in both land and ocean flora and fauna in the period (millions of years) leading up to the K/T event. Hallam (2004) has called the bolide impact just the 'final *coup de grace*'.

At least one other site, the Shiva crater off the west coast of India, dates to the same time of 65 million years ago. It is quite possible that there were two or more impacts at that time responsible for the K/T event. They may have originated from a single body which disintegrated before reaching Earth (like the 1994 Jupiter event discussed in Chpater 5).

Conclusions

In the course of its 3.7 billion year history life on Earth has undergone numerous highly stressful periods and events. Even so, it was never totally extinguished. Life has been (and is) remarkably resilient and has always recovered. The very changes in the environment were the driving factors of evolution that produced species better adapted to the new situation. Moreover, convergent evolution often leads to other species taking over, where the first may have gone extinct (Martin and Meehan 2005).

Despite this resilience especially at the microbial level, most species which have ever emerged on Earth have succumbed to the stresses of the biosphere. The more complex the life form, the more difficult it is for the species to adapt. Consequently, multicellular species have rarely persisted for more than about a million years before going extinct.

Very lucid summaries of what is known of extinctions and their causes can be found in the books by Courtillot (1999) and Hallam (2004). The relationship between Phanerozoic extinctions and plate tectonics has been reviewed by MacLeod (2003). Brenchley and Harper (1998) give an extensive treatment of the methodology of fossil study in relation to palaeo-ecosystems.

Summary

- Ever since life appeared on Earth about 3.7 Gy BP it has had to contend with an ever changing, often highly stressful, environment.
- Earth's second atmosphere evolved into the third, oxygen-rich atmosphere. Photosynthesis produced the oxygen and lowered CO_2 levels. However, a change in atmospheric oxygen and CO_2 could only happen when the carbon products were sequestered by an overall increase in plant material (e.g. forests), subduction by movement of tectonic plates, and precipitation to the sea floor of calcium carbonates of plant origin and silicon-containing diatoms (the latter in the last half billion years).
- Earth's third atmosphere, as it is today, contains some 80% v/v nitrogen, 21% oxygen, and 0.028% CO_2 (pre-industrial values). It also contains from 0.1–3.0% water vapour. Nitrogen, essential for all life forms, can only be used directly by a few nitrifying bacteria and some cyanobacteria. Multicellular plants use forms of nitrogen derived from microbial nitrifiers. Heterotrophic life forms obtain nitrogen by consuming other life forms. Today's high levels of oxygen inhibit nitrification, severely stressing multicellular life.
- The methane of the early atmosphere was oxidized by oxygen at the same time as the level of CO_2 was falling. Both methane and CO_2 are greenhouse gases. This was one of the main causes of the first ice ages, which were so severe as to affect *almost* the entire surface of the Earth ('Snowball Earth').
- Respiration, which releases energy stored in molecules derived directly or indirectly from plants, was initially anaerobic. It is relatively inefficient: for example only 2 ATP are produced from each mol of glucose. The high oxygen content of the third atmosphere enabled aerobic respiration, which produces 38 ATP from the same amount of glucose. Aerobic respiration facilitated the evolution of multicellular life forms such as insects and vertebrates, which require much rapidly supplied energy.
- Occasional very high oxygen levels, which could reach 35% v/v, caused worldwide fires and the formation of coal strata in the Carboniferous period and inhibited plant growth prior to the K/T extinction event.
- High oxygen and low CO_2 inhibit photosynthesis in most plants to this day. This is mainly a result of the affinity for oxygen of Rubisco, the enzyme for carbon fixation, and its relatively low affinity for CO_2. High oxygen levels also allow the formation of very reactive and toxic oxygen radicals.
- Apart from the early, protracted and severe ice ages, the Earth has passed through shorter, recurring ice ages, especially in the last half million years. The cause(s) are uncertain, but include alterations in the composition of the atmosphere, cyclic changes in the tilt of Earth, and periodic passage of the Sun through the Orion

arm of the Milky Way. The latter modulates the cosmic ray radiation reaching Earth, which affects cloud formation. Earth is currently emerging from the most recent ice age, which ended 11,000 y BP.

- The fossil record of the Palaeozoic era indicates that life has gone through five major and some 16 lesser extinction events. The greatest was at the end Permian (225 My BP) when 75% of land and 85% of all marine species perished. There are varied and possibly multiple causes of these calamities: extraterrestrial, such as impacts of comets and meteorites, or radiation surges from nearby supernovae; and terrestrial, such as ice ages, massive volcanic eruptions, and large scale out-gassing of hydrogen sulphide from oxygen-depleted oceans. The K/T boundary extinction was certainly caused by a bolide impact but, even before the impact, the biota had begun to deteriorate as a result of extreme volcanism, high atmospheric oxygen, and possibly other terrestrial factors.

- Earth today is still being struck by comets and meteorites, such as the meteorite which fell in what is now Arizona 50,000 y BP or the small comet, or asteroid, which vaporized over Tunguska, Siberia, just 100 years ago.

- Life in general, especially at the microbial level, has been amazingly resilient, always adapting to the new biosphere environment. Contrariwise, most multicellular species did not survive for more than 1–3 million years.

- Species' longevity decreases as life forms become more complex.

8

Mechanisms of evolution: from first cells and extremophiles to complex life

The dating problem

We learn of former life from its signatures (fossils and organic and isotope tracers) which are found in the different geological strata of the Earth, and from the physiology, biochemistry, embryology, and genetics of early, but still extant, life forms. It would be easy if the deeper we dug the older were the strata. In a general way this is true, but upheavals and subsidence often confuse such a neat picture. This greatly troubled the first palaeobiologists striving to determine the dating and order of appearance and disappearance of species. It led to never-ending disputes as to the age and periods of life on Earth. Dating arguments sometimes became circuitous; with the age of a newly found stratum being based on the fossils it contained and vice versa.

These debates came to an end with the development of isotopic dating methods (see part 2 of Lunine, 1999). Radioactive isotopes have constant exponential rates of decay, typical of each element. They radiate energy while decaying to a lighter, usually stable element. Decay begins immediately following the formation of the radioactive isotope from another, stable element. The radioactive isotope forms in the atmosphere or on the surface of the Earth, as a result of cosmic radiation. For example, when ^{14}N in the upper atmosphere absorbs neutrons from cosmic rays, ^{14}C is formed. ^{14}C is unstable and breaks down, emitting beta rays and reverting to ^{14}N. Formation of ^{14}C stops from the moment ^{14}N is protected from cosmic radiation by the lower atmosphere or by the surface material, soil, or rock. The percentage of the original ^{14}C isotope remaining is an indication of the time during which the isotope was isolated in the strata. Radioactive carbon (^{14}C) has a half-life[1] of 5730 years and is used for dating relatively young, organic material. Other isotopes with greater longevity, such as potassium (^{40}K) which has a half-life of 1.25 Gy, are used to date older strata. There are many problems with the use of isotopes for dating, such as heating which can drive out the degradation products, but they can usually be resolved.

---

[1] The time it takes for half the atoms of a radioactive isotope to undergo radioactive decay.

Timing of the main events in evolution

It had long been suspected that biological species evolve from previous, simpler species. The terms 'evolve' and 'simpler' need some explanation. In purely evolutionary terms a successful species is one that replicates easily and adapts to its environment, which includes competing with other species. With this definition there is no doubt that 'simple' bacterial species are far superior to more complex life forms, such as mammals. Insects and other animals, with their fast metabolism, which enables motion and the use (by humans at least) of manual tools, and their nervous systems with environmental sensors and brains, are at the 'top' of the evolutionary ladder.

In recent years genealogy has benefited from molecular biology. It is now possible to analyse and follow the appearance of gene structure by sequencing DNA nucleotides thus clarifying evolutionary pathways. With some exceptions, the results of these new methods have shown at least qualitative and sometimes quantitative agreement with the older dating estimates. The integration of these studies, classical and modern, have produced the time line of geological strata and evolution shown in Table 8.1.

Some early ideas about evolution

The Hebrew Bible, with its six days (periods?) of creation, culminating with humans, is perhaps the most well known (and misquoted[2]) source. There were many other myths of ancient peoples, all fascinated with the question of our antecedents. Perhaps the most interesting was Anaximander of Greece, in the 6th-century BCE. He posited, correctly, that life was of aquatic origin.

One of the first scientists to develop the idea of progressive evolution from one species to another was the French zoologist J. P. Lamarck (1744–1829). He believed that exposure to environmental factors brought about adaptations which were inherited by the next generation, slowly 'improving' the species. A common example of such thinking is the extension of the necks of giraffes. Giraffes' necks were thought to have grown a little longer in each generation, as the animals strove to reach ever higher tree leaves. In the early 20th century the concept of the inheritance of acquired characteristics was shown, empirically, to be incorrect. This did not prevent the adoption of Lamarckism by Trofim Lysenko, in mid 20th-century Russia. In 1928 he claimed to be able to impart cold resistance from one cold-treated generation of wheat to the next. The concept fitted well with the dialectic ideology of Soviet communism. It dominated and retarded Russian biology for half a century. This should not detract from our appreciation of Lamarck, who understood the principle if not the mechanism of evolution in the early 19th century.

In the first half of the 19th century, two field biologists, Alfred Russell Wallace and Charles Darwin, quite independently and using data sets they had acquired in very

---

[2] Misquoted, as the Bible, whether or not divinely inspired, is not a book of history, science, or evolution.

Table 8.1 Major geological and evolutionary periods and events on Earth

Aeon	Era	Period	Date (My BP)[a]	Major life forms or occurrence
Hadean			4550	Formation of Earth
Archean			4550–2500	First Earth crust
				Late heavy bombardment
				First life ~3700 My BP (LUCA)[b]
				Prokaryotes, chemoautotrophs
				First photosynthetic cells
				Moon still causing 300 m high tides, with high frequency
Proterozoic			2500–542	First eukaryotes
				First simple animals
				Sexual reproduction 850–630 My BP; global glaciation ~550 My BP; UV blocking, oxygen- and ozone-containing atmosphere
			~530	Cambrian 'explosion'
Phanerozoic	Palaeozoic	Cambrian	542–488[c]	Arthropods
		Ordovician	488–438	~475 My BP first land plants
		Silurian	438–408	Marine invertebrates
		Devonian	408–360	Ferns
		Carboniferous	360–286	First seed plants
				First insects
		Permian	286–245	First reptiles
			251	Greatest extinction event (95% of all species)
	Mesozoic	Triassic	245–208	Reptiles
				Rich marine life
				First mammals
				Gymnosperm forests
		Jurassic	208–144	Major dinosaur period
		Cretaceous	144–66.4	Birds
				~130 My BP- Angiosperms (flowering plants)
			65.5	Extinction of 50% of species (including all non-avian dinosaurs)
	Cenozoic	Tertiary	66.4–1.6	Birds and mammals Grasslands
				First C_4 plants (?)
		Quaternary	1.6–0.2	*Homo erectus*
		Holocene epoch	0.2–Present	*Homo sapiens sapiens* (modern humans)

[a]My = million years. These dates are estimates. They vary, depending on different definitions of what defines a Period. The evolutionary periods they separate and define were given either descriptive names, such as the 'Phanerozoic' meaning 'conspicuous animal life' or names of the regions where certain fossil strata were first identified, e.g. Jurassic from the 'Jura', a mountainous region between Switzerland and France.

[b]LUCA – Last universal common ancestor.

[c]Note that few geological times can be dated with great accuracy. Different sources give Cambrian events which differ by ±2 My.

different parts of the world, developed the hypothesis of evolution by way of natural selection. According to their concept, new traits appear at random. A very small percentage produces some competitive advantage. When this happens, a new species is formed, better able to compete with others. Darwin published his major work *On the Origin of Species by Means of Natural Selection* in 1859, after publishing his earlier findings and conclusions together with Wallace (see Reference and Resources for details of recent editions). Of the two, Wallace had the wider interests, which covered politics, literature, and, unfortunately, spiritualism.

The definition of 'species' is difficult. For most biologists, it means a line of life (insect, animal, or plant) which does not interbreed with other lines (Mayr 2004). It is sometimes used loosely to define a line with a new trait, which may interbreed. In this case it should be called a 'variety'. This meaning of species cannot be used for unicellular organisms and asexually reproducing advanced life. Today, prokaryotes are considered as belonging to the same species if 70% of their DNA sequences are similar. In the past, similarity of form and function has been used, not very satisfactorily, for a species definition of microbes. The problem of defining 'species' has remained to this day (Zimmer 2008). Darwin recognized, but avoided trying to resolve this difficulty, whether in prokaryotes or advanced complex eukaryotes.

The work of Darwin established evolution by natural selection as the central paradigm of biology. It has lasted from the second half of the 19th century to this day. Darwin did not offer a clear concept of the mechanism of evolution. In fact, unlike Wallace, he did not entirely rule out Lamarckian inheritance of acquired characteristics. Darwin put geological gradualism (as opposed to catastrophism—see below), competition between species, and steady, incremental changes, at the centre of his theory. In his later papers he sometimes used the expression 'survival of the fittest', often quoted as being the central tenet of Darwinism. This phrase was in fact the invention of his friend, the philosopher Herbert Spencer. It actually says little, as 'fittest' is defined by survival. Darwin's backing of gradualism, his occasional support of the concept of the inheritance of acquired characteristics, and his use of poorly defined terms, do not significantly detract from the scientific edifice he created. The synthesis of our present understanding of evolution has produced the evolutionary tree depicted in Fig. 8.1.

The overall picture of evolution shown in Fig. 8.1 is clear, although doubts remain as to the placement of certain species, branches, and lines. Evolution certainly does not always proceed neatly along clear lines. There are numerous examples of parallel evolution and convergence and of lateral, as compared with vertical, evolution. Moreover, as discussed above, all this evolutionary 'progress' was occurring under constantly changing and often very stressful conditions, which were often the drivers of evolution.

Four major events are recognized which determined the course of evolution of life on Earth, and numerous lesser ones. The more important can be seen in Table 8.1 and Fig. 8.1. They include: the end of the late heavy bombardment at the beginning of the Archean aeon, which enabled the onset of life; the Cambrian 'explosion' which ended the Proterozoic aeon and was characterized by a huge increase in the number

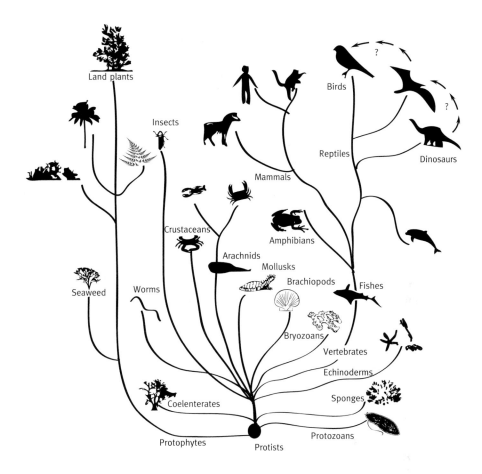

Fig. 8.1 The phylogenetic tree of multi-cellular life.

and sophistication of multicellular organisms; the extinction event at the end of the Permian, which almost ended life 251 My BP and the extinction at the end of the Cretaceous, 65.5 My BP, which erased 50% of all species, including all the non-avian dinosaurs. The latter event enabled the proliferation of birds, mammals, and other modern species. (See below for further discussion of the above two extinction events.) As for the 'Cambrian explosion', there certainly was a huge increase in multicellular organisms at that time, from which all present, advanced complex life developed. However, it is no longer considered to be quite so dramatic, time wise, as originally thought. Some multicellular life forms have been traced to the late Proterozoic aeon, pre-dating the onset of the Cambrian by some tens of millions of years. Even so, there was a tremendous and relatively rapid burst in evolution, after some 2 G years of uni-cellular, mainly prokaryotic life. The reason for this enormously important biological advance after so long a period of evolutionary stasis is unknown. The long period of relative stasis (2 Gy) is of great importance when we try to calculate the chances of finding other advanced life in the universe: 2 G years is one-sixth the life of the universe since the Big Bang.

The beginning of the 20th century saw the development of quantitative Mendelian genetics.[3] At first it was thought that genetics and Darwinian evolution were incompatible. Later work in the 1930s showed that they could be reconciled, especially when field studies of population dynamics were taken into account. This integration received the name 'neo-Darwinism' or the 'modern synthesis' (Huxley 1942). As discussed below, while modern molecular studies have added much understanding to the cellular basis of evolution, the mode of operation under field conditions still raises many questions. Three other major contributors to the 'modern synthesis' who gave it a quantitative background were Ronald Fisher (the father of all modern biological statistics), Theodosius Dobzhansky, and John Maynard Smith (see Smith, 1982, for a summary of his and their work).

At this point it should be noted that although there are sometimes passionate disagreements among evolutionists, the basic concepts are accepted by all. Darwinian evolution is considered to have been established by innumerable supportive findings (with the usual reservation that facts in science can change!). It is no longer considered to be a 'theory' or 'hypothesis'. Disagreements exist, but they are in the realm of operational mechanisms.

Evolution is basically a result of the churning, stochastic interaction of many billion separate entities. Moreover, the individual life forms are constantly undergoing modifications. This may be the result of chance mutations, symbiosis, reshuffling of existing, previously neutral genes, or new modes of activation and combinations of existing genes (see below). However, evolution is not entirely stochastic, as it is driven by the 'rules' of natural selection. New characteristics survive when they improve fecundity, adaptation to the environment, or competitiveness.

Consider weather, a truly stochastic[4] process limited only by the laws of chemistry and physics. This is a process with which we are all familiar, on a daily basis. It is produced by the interactions of millions of packets (or partially discrete masses) of the atmosphere, each with a certain gas and water vapour content, temperature, pressure, etc. The air masses swirl around, mixing and interacting with each other, exchanging mass, momentum, energy, and compositional content. In addition they may be modified by contact with warm or cold seas and land masses as they rise or sink in their forced passage over oceans, lakes, and mountains. At the same time they receive

[3] Gregor Mendel was a 19th-century Austrian monk whose garden experiments on the inheritance of plant characteristics (e.g. colour, size, etc. of peas) set the basis of modern quantitative genetics. His work was at first overlooked. It was only rediscovered and recognized for its great contribution at the beginning of the 20th century.

[4] Apparently purely random.

changing amounts of solar radiation. Given the resulting turbulent motion and energy exchange, governed by just a few 'laws and generalizations' (such as global movement of air streams) meteorologists have great difficulty in predicting local weather. This is so despite their having available thousands of physical measurements of the above parameters. These data are fed into sophisticated mathematical models of global circulation, for solution on supercomputers. The end results are weather predictions which, despite these advanced methods, are reliable for at most a few days.

Very similar reasoning applies when trying to understand the vagaries of evolution, over aeons of time; but evolution of life and the prediction of its future course are much more complicated. The physicochemical characteristics of an air mass are many orders of magnitude fewer and simpler than those of the most basic living cell. Given this turmoil, it is not surprising that there are numerous, often conflicting theories as to how some of the results of evolution (Fig. 8.1) came about. Predicting the future course of evolution is equally impossible.

Some recent theories of the mechanism of evolution

Two of the most influential late 20th-century evolutionists were Richard Dawkins and the late Stephen Jay Gould. Their often entertaining, but always erudite, exchanges were famous. Their disputes on mechanisms have often been misquoted by Creationists as evidence that even scientists have problems with evolution. But this is not so. Gould and Dawkins were/are both fervent 'Darwinists' and it is often difficult for the non-expert to understand where exactly they differed. To venture an oversimplification: Dawkins saw the well-being of the gene as the main driver of evolution while Gould looked more towards the whole body and population. A review of their work is given by Sterelny (2001), whose book, published just before Gould's death in 2002, carries the germane title *Dawkins vs. Gould: Survival of the Fittest*.

Game theory and evolution

Evolution is hardly a game but it does operate in ways which are often comparable to one (the classic work in this field is that of Smith, 1982). In games there is an intention to achieve a certain outcome. For example in bridge we want to get the most points, or in betting the most cash. These are wilful intentions. As noted above, evolution preferentially selects certain characteristics which together favour survival. The difference here is that the driver is purely passive. Natural selection is not wilful. Even so the strategies selected by evolution may be similar to a game and are therefore amenable to analysis by the same branch of mathematics, game theory, which is being applied to many real-life situations, such as investments in shares (Cohen and Mangel 1999).

One example from the world of plants can illustrate this approach: the strategy of seed germination (Cohen 1967). When a farmer buys seeds he is interested in getting close to 100% germination, soon after sowing. The situation in the wild is quite different. At certain seasons of the year seeds are subjected to conditions which enable germination-mainly water and a warm temperature. It often happens that after germination and the production of young sprouts, conditions become inclement to growth and the seedling dies. This could be because of flooding, drought, cold, or many other environmental stress factors. If a plant species had 100% germination it would rapidly be eliminated, even if such sprout-lethal conditions occur only once in many years. On the other hand, if they leave part of their seeds dormant to the next season, as a safety measure, they run the risk of they're being eaten by insects, birds or other herbivores. Consequently the plants must develop survival 'game' strategies. They have to solve the problem of what percentage of seeds should be left dormant to the next and following years to ensure long-term survival over many years. The solution must take into consideration the statistics of weather, climate, and seed-eating grazers (Cohen 1967; Evans *et al.* 2007).

Under desert conditions, for example, the uncertainties of weather are always great. It has been shown that in arid lands a larger percentage of seeds are left dormant than in plants growing in the tropics, where environmental conditions are much more stable. Plants do not of course 'decide' what to do, but the statistics of natural selection drive the survival of those species in which the division between germination and dormancy best fits the region where they are found.

Punctuated equilibrium

Eldredge and Gould (1972) published a seminal paper on the palaeobiology of animals. In this work they concluded that evolution proceeded not by gradual increments, as in classical Darwinism, but rather in stops and starts. Their work was based on fossil evidence from many animal lines, from snails to horses, found in well-dated strata. They reported that evolution went through long periods (millions of years) of stasis, or equilibrium, and then suddenly started again. This has since been supported by many other studies, in which little evidence was found for Darwin's slow, gradual evolution. Eldredge and Gould had no satisfactory explanation for this behaviour. Gould proposed one possibility: that a new species would evolve when a population of the species is isolated geographically. For example, this could happen on a newly formed island. One criticism of punctuated equilibrium, as this theory became known, has been that a dearth of fossil evidence could give a false impression of long periods of stasis, whereas in fact there may have existed undiscovered intermediate species ('missing links'). Supporters of punctuated equilibrium believe that the shear mass of evidence, from various study sites, makes this objection unacceptable.

Evolution and symbiosis

Lynn Margulis published her first paper on endosymbiosis and the origin of eukaryotic cells in 1966 (see Margulis and Sagan, 1995, for a later synthesis). It included the then outrageous idea that mitochondria and chloroplast plastids arrived in eukaryote cells as a result of the invasion by prokaryotes. Mitochondria are cell organelles which carry most of the enzymes for respiration. They are found in all eukaryotic and prokaryotic respiring cells. Chloroplasts are organelles which contain the photosynthetic pigments and mechanisms for radiation and carbon capture. They are found in prokaryotic algae (especially the intermediate cyanobacteria) and in the leaf cells of all photosynthesizing higher plants. According to Margulis, the donor prokaryotes of this symbiotic relationship had already developed the mechanisms for respiration and photosynthesis within their cytosols. Mitochondria can only arise by fission of existing mitochondria, which further supports their independent origin. A similar idea to that of Margulis had been proposed by the Russian biologist Konstantin Mereschkovsky in 1905, but his work was overlooked.

With the development of molecular biology in the 1980s, Margulis' ideas were vindicated. Both types of organelles were found to contain genetic material independent of and largely different from that of the nucleus of the host, but close to that found in prokaryotic cells. Recent studies have shown that, in the course of evolution, the genetic material of the invading mitochondria and chloroplasts becomes diluted by that of the host nucleus. Today the plastids of the eukaryotic cyanobacteria have only 5–10% of the genes of the putative prokaryotic source, while the nuclei contain much DNA of cyanobacterial origin.

The concept of symbiosis is well known in biology. A common example, mentioned previously, is that which exists between *Rhizobium* bacteria and leguminous plants. In this case the bacteria supply inorganic nitrogen to the plant while the plant supplies energy-rich nutrients and a low-oxygen environment to the anaerobic bacteria. Although there are many such examples to be found in biology, symbiosis was considered the exception and not the general case.

Margulis extended her ideas and built an entire critique of neo-Darwinism, with philosophical and political overtones. According to Margulis, the main driving force for evolution is not competition, as in classical Darwinian evolution, but rather cooperation.[5] Margulis' scientific findings and main concepts are accepted today as explaining many evolutionary findings. Her support of fringe ideas, such as the Gaian hypothesis, has not always gone down well with her colleagues.

The Gaian hypothesis was proposed by James Lovelock in 1979. It recognized that the biosphere of Gaia (a.k.a. Earth) has many feedback mechanisms. It proposed that the Earth behaved *as if* it were a single organism, controlling its homeostatic equilibrium, which is now being perturbed by humans. He gathered many supporters, many with a mystical bent, who omit Lovelock's 'as if' caveat, and discuss 'sentient Gaia'.

---

[5] No doubt Margulis, like many others, was much irked by the 20th-century Nazi adoption, and unwarranted application to the human situation, of classical Darwinism. The original 1859 edition of Darwin's *magnum opus* 'On the Origin of Species by Means of Natural Selection' carried an ominous and easily misinterpreted sub-title 'Or—the preservation of favoured races in the struggle for life'. Later editions omitted this phrase. This fitted all too well the 'superman' ideas of the 19th-century philosopher, Nietzsche, another favourite of Nazi ideology. Margulis sees cooperation, not competition, as being the central driver of evolution.

It has, however, inspired the discovery of many bona fide global, ecological feedback mechanisms (Volk 1998).

Neutral theory of molecular evolution

This hypothesis was first put forward by Kimura in 1983 (see Kimura 2008). In classical Darwinian evolution it is assumed that most mutations are deleterious and are rapidly eliminated. Only the very rare advantageous mutations survive, sometimes forming new species. In Kimura's view most mutations are neutral; they are neither harmful nor beneficial. They do not disappear but accumulate. Eventually selection operates via the entire store of alleles.[6] There may be many alleles which produce equally competent solutions. There are some beneficial, evolution-driving mutations, but most selection is the result of genetic drift and selection acting on otherwise neutral, almost equally valuable, alleles or groups of cooperating alleles.

Although most of the concepts of the neutral theory are now accepted, there are still problems with its general applicability. It has proven to be a very powerful basis for molecular clocks, with which it is possible to calculate the time passed since given species diverged from a common ancestor. It also fits well with the latest ideas of 'evo-devo' described below. However, its predictions have sometimes failed to account for field distributions of diversity (cf. Dornelas *et al.*, 2006, working with corals).

Complexity theory

This is a rather new, mathematically based approach to understanding the behaviour of systems containing numerous essentially stochastically interacting entities. Examples are the stock market, weather, or evolution. Processes in these systems do not proceed along lines which can be described by normal linear mathematics. Sometimes limits are reached from which there is a switch in the situation, requiring a new algorithm. For example, a generally favourable economic situation and wildly euphoric investors may drive up the price of shares at a steady daily rate until the 'bubble' bursts and the shares crash. The analysis of such nonlinear systems has been given the name 'complexity theory'. Chaos theory is a branch of complexity theory which deals with situations where there are turbulent interactions of numerous entities, but only a few species. This is the case with weather, or the now rather familiar fractals. Chaos theory is successful in describing the formation of irregular bodies, such as cloud shapes, shorelines, and plant morphology.

Complexity theory indicates that in large-scale stochastic systems both the formation and the breakdown of structure are predicted to occur, even without the intervention of major outside factors. (Energy exchange may be required to enable increase or decrease of entropy.) This approach may explain bursts of evolution (a 'saltation' in the jargon of complexity theory) which come after long periods of

---

[6] Alleles are different operational forms of the same gene. For example, two alleles of the same gene governing hair colour may produce either brown or blonde hair.

evolutionary stasis. An example of this is the Cambrian explosion which came after the long Proterozoic period of evolutionary stasis. It also supports the general hypothesis of punctuated equilibrium. Obviously, forces from outside the biosphere have affected life, sometimes very dramatically. The basically statistical approach of complexity theory can therefore only be an additional consideration in our understanding of the actual, highly erratic course of evolution, during the 3.7 Gy of life on Earth. Many relevant papers introducing complexity theory can be found in Crutchfield and Schuster (2003). A less mathematically demanding review of life and complexity has been given by Zimmer (1999).

The continuing coevolution of molecular biology and evolutionary theory

Although Darwinian evolution is now well accepted, the mechanisms of its operation, as seen in the fossil record, still pose many questions and encourage numerous theories, such as those briefly described above.

After the discovery of the structure of DNA by Watson and Crick, in 1953, and the subsequent rapid decryption of the genetic code, optimism was rampant that the entire controlling system of biology and its evolutionary mechanisms would soon be understood. Only a few biologists, such as Richard Lewontin (see Lewontin, 2000, for a collection of his prescient essays), seemed to realize the daunting complexity in going from genes to operating proteins and differentiated tissues and organelles. The field is currently very dynamic, struggling with the problems foretold by Lewontin. With each new breakthrough in our level of understanding a new subfield emerges, often given a new soubriquet, as in the above section subtitle.

The human genome (the DNA coded sequence of genes) contains some 24,000 genes. This is not a large number compared, for example, with that of wheat which contains about double this number. Once it was thought that each gene coded for a single protein. But in humans, for example, the number of proteins is in excess of 250,000. Today it is recognized that the interspaces between the genes are filled not by 'junk' or useless vestige DNA, as originally thought, but by DNA which can activate or deactivate combinations of genes and proteins. The study of the entire genome and the interaction between genes is known as 'genomics'. It was soon realized that the groupings and expression of proteins produced via DNA coding and RNA intermediaries are never constant. They vary in different tissues and certainly in different species, and are modified by the environment. Moreover, these changing protein groups also have a reverse effect, affecting the activation or suppression of genes[7]. The study at this level has been called 'proteomics'.

One of the major problems of biology is differentiation. Its study pre-dates molecular biology by at least a century. It began even earlier as 'embryology', the study of the

---

[7] Note that today's understanding is contrary to the original basic 'dogma' of molecular biology, which postulated that cell control was unidirectional, from DNA via RNA to proteins.

development and differentiation of embryos. Whereas each cell of a multicellular body has a full complement of genes, capable of producing any part of the body, it is obvious that only part of the genome is expressed (no arms sprouting from heads or eyes from legs). Moreover, on an interspecies level, many different species (e.g. cows and humans) have almost the same genetic DNA, despite the obvious differences.

The control of differentiation is now known to be in the hands of a small number of special genes, the so-called 'HOX' genes. They are able to activate, or deactivate, genes and groups of genes from the entire library of the genome and the intergenic DNA. In this way they are active in both evolution (by selecting a new, more adapted combination of gene and protein groups) and in the specification of the genetic material to be used for a specific bodily organ or tissue. Note that this form of adaptation and 'selection of the fittest' is carried out without necessarily being based on new mutations at the DNA level, although the latter certainly occur.

An enigmatic fact (which has been noted for at least two centuries) is that during their development, many embryos go through stages which appear to recapitulate the evolution of the species. For example snake embryos often have rudimentary legs; early mammalian embryos are fishlike. The study of the development of embryos and their later differentiation and specialization may shed light on their prior evolution. This entire field, including the study of the control and outcome of HOX gene and intergenic DNA activation, has been given the name 'Evo-Devo' (evolutionary developmental biology). More detailed presentations of Evo-Devo can be found in Prud'homme *et al.* (2007) and the review by Carroll *et al.* (2008).

Further readings in general evolution

A general description of the course of evolution, as shown in Table 8.1, is outside the scope of this book, apart from a number of important aspects which illustrate the cosmic and geological connections. There is no shortage of textbooks and popular writings on evolution (Google lists about 1.76 million entries for 'evolution textbook'!). It is consequently not easy to recommend books on evolution, especially at the less technical level.

An already rather old but classic popular discussion of the importance of genes, reproduction, and competition in evolution is that of Dawkins (1991). Early evolution is covered in Fenchel (2002) and Knoll (2003); that of more advanced animals is described in Gould (1990) with a very comprehensive treatment of evolution (mainly animal) in Gould (2002). Evolution of plants, with emphasis on their effect on the atmosphere, is covered by Beerling (2007). Many important discussions of the changing physical environment of Earth as it affected the evolution of life can be found in Rothschild and Lister (2003). Two very readable reviews of what is known of evolution and of evolutionary mechanisms can be found in Fortey *et al.* (1997) and in Carroll *et al.* (2008), respectively. Readers wishing to gain a background in the basic biological and evolutionary thinking of the 20th century will enjoy Ernst Mayr's (2004) collection of essays (the most recent were written close to his 100th birthday!). The compilation on complexity theory by

Crutchfield and Schuster (2003), mentioned above, is recommended for those with a mathematical background.

Extremophiles

Extremophiles, as their name implies, live in the most unusual parts of the biosphere, often under apparently stressful conditions. 'Apparently', as for many of these organisms the environmental niche in which they live is not stressful for them. They are adapted to these extremes and benefit from having fewer competitors.

Astrobiologists are interested in extremophiles because conditions similar to those in which Earth extremophiles live have been found in a number of sites within the Solar System (Chapter 4). Their very existence, and the existence of these extraterrestrial life possible niches in the Solar System, increase the likelihood of finding life, at least on the unicellular level, in extra-Solar System planets. The latter are only now entering the range of our Earth-bound sensors. It is expected that in the next decade we shall learn far more of the environments of some of these planets (Chapter 10). Just recently, methane and water were detected in the atmosphere of a planet circling a star at a distance of 63 ly from Earth (Swain *et al.* 2008). Although suggestive of life, and very exciting, these molecules were probably of non-organic origin, as the surface temperature was calculated to be too high for liquid water.

Extremophiles are of interest to astrobiologists studying Earth, as some of the niches where they live are not unlike what is believed to have existed during the Archean aeon when life began. Many papers on unicellular extremophiles can be found in Seckbach (2006). A more popular review of extremophiles is provided by Gross (1997). The hypothesis paper by Cleaves and Chalmers (2004) provides a short general review of extremophiles in relation to the chances of finding extraterrestrial life. They take a rather pessimistic position, contrary to that of Rothschild and Mancinelli (2001). An exposition of the hidden biosphere, deep in the Earth, is given in Gold (1999). His ideas are often controversial (see below) but always stimulating.

Adaptive mechanisms often come at a price, in terms of allocation of resources such as energy and carbon. These requirements often, but not always, limit the ability of such organisms to spread into other environments. The 'phile' at the end of extremophile means 'love', but is sometimes an exaggeration. The adaptation may be facultative not obligate. For example, the halophytic plant *Atriplex halimus* grows in arid, salty deserts as a low bush, with small leaves, reaching a height of about 1 m. When *A. halimus* grows in a mild climate, with fertile soil and a good supply of water, it can reach 4–5 m, with large leaves, flowers and seeds.[8] This is an example of phenotypic plasticity, a significant evolutionary advantage.

There are many examples of extremophiles living under conditions that are severely inclement for most other life forms on Earth, but this is not entirely unexpected. Even

---

[8] Unpublished observation of the author.

in a 'normal' homogeneous population the distribution of growth parameters results in some 4% of the population falling more than two standard deviations from the mean. But extremophiles also have mechanisms which place them outside the 'normal' population.

The special adaptive mechanisms of extremophiles are sometimes of practical interest. Animal and plant breeders seek to transfer their stress-resistant characteristics to normal agricultural varieties. For example, the salt resistance of low-yielding halophytic tomatoes has been transplanted to high-yielding salt-sensitive agricultural varieties. Today, many enzymes, antibiotics, and other biomolecules derived from extremophiles are being used in industrial processes (see Table 3 in Rothschild and Mancinelli, 2001).

Again it must be repeated that our search for conditions supporting extraterrestrial life is coloured by what we know of Earth. The conditions under which the most extreme of Earth extremophiles exist represent only a very small fraction of what is possible on other planets. For example, based upon what is known of sulphur bacteria growing near volcanic springs on Earth, Schulze-Makuch *et al.* (2004) have speculated on the possibility of life existing in the acidic Venusian atmosphere. Chyba and Philips (2001) discussed the possibility of life below the ice covering the liquid ocean of Europa (Chapter 4) with reference to what is known of life in the oceans below the ice in the polar regions of Earth. But Cleaves and Chalmers (2004) have pointed out that even where there are conditions similar to those supporting life on Earth, life may not necessarily begin spontaneously.

Some examples of extremophile behaviour

Resistance to drought

We tend to think that most extremophiles are prokaryotes, but this is not always so. Adaptation to stress extends right through the evolutionary tree and is indeed the central driving force of evolution. So extremophiles are to be expected, and are found, at all levels. Some species of desert mice tackle the problem of water shortage by husbanding. They obtain much of their water from almost dry seeds and plants and excrete very dry urine and faeces. Frequently an organism will show adaptations to more than one stress. Wong *et al.* (2005) describe a plant adapted to cold, drought, and salinity, each adaptation being controlled by a different gene or group of genes.

Many organisms adapt to stress by avoidance. Bacteria sometimes have dormant forms which can survive desiccation for long periods (during which they neither grow nor reproduce). Dormant plant seeds are similar to the bacteria in this. When an organism combines desiccation resistance with low susceptibility to radiation, it becomes a prime candidate for panspermia (see Chapter 6).

Some desert plants are able to pass through extremely dry periods while conserving their leaves and photosynthesizing mechanisms, which return to full activity shortly after a period of rainfall. The most well known of these is the south-west African gymnosperm *Welwitschia mirabilis* (the 'resurrection plant'; Fig. 8.2). Its drought- and heat-resistant characteristics enable individuals to survive in the most arid deserts for as long as 1000 years or more. It is currently being saved from extinction, in Angola by the presence of innumerable land mines. (Is this an example of the benefit of human activity to the biosphere?!)

Radiation resistance

As discussed above and in Chapter 6 (see section on panspermia) this is an ability which may be of importance to organisms travelling through space (if there are any such). As noted, *Deinococcus radiodurans* is probably the champion in the field of radiation resistance having been found (alive!) within the central cores of decommissioned nuclear power stations. The ability to resist or to recover from radiation damage would have been advantageous in early Earth, before the high-oxygen/ozone atmosphere provided a shield against solar UV radiation (Chapter 7).

Resistance to extreme temperatures

Extreme temperatures, both high and low, present a challenge for life on Earth, quite apart from the 0–100°C range (at 1 bar pressure) requirement for liquid-phase water. Enzymes usually operate optimally at temperatures close to those prevalent in the normal environment of their species. There are hyperthermophilic organisms which grow at temperatures up to 100°C, a level at which most enzymes denature (Rousell *et al.* 2008). They contain special heat-stabile proteins. Bacteria grow around undersea vents at temperatures above 120°C. Although water at these depths remains liquid, as a result of the high pressure (increasing by 1 bar for every 10 m depth), they still face the problem of stabilizing their proteins. It has been suggested that the universal ancestor

Fig. 8.2 Welwitschia mirabilis, the 'resurrection plant', growing in the Namibian desert. Courtesy of Wikipedia.

of life should be sought among the thermophiles (Di Giulio 2003), but there are other candidates (Bada and Lazcano 2002).

Psychrophiles grow at temperatures below 15°C. Some even grow below 0°C, if the salinity is sufficient to prevent water freezing. They are the ones most relevant in the future search for life in the sub-ice ocean of Europa.

Resistance to acidity and alkalinity

Earth organisms usually have a very small tolerance of change in acidity (pH). They tend to live in environments either close to pH 8.1–8.2 (sea water) or 7.0–7.2 (fresh water). A very few are able to grow under extremes of acidity or alkalinity. The red alga *Cyanidium caldarium* has a pH optimum of ~2.5 but will survive 0.5. Other bacteria and some fungi have been found growing in soils with a pH < 1.0. Although a high pH tends to inhibit the handling of energy (especially the movement of protons through membranes) many eukaryotes, even multicellular organisms including some fungi, have been found growing at pH > 10. For a molecular-level analysis of acidity tolerance see Chi *et al.* (2007).

Ecosystems around deep sea vents

The importance of deep sea vents has already been mentioned above (see also Beatty *et al.* 2005). They are found around deep sea trenches, where there is venting of hot gases from within the Earth. Numerous such vents have been found, surrounded by entire independent ecosystems. Their primary energy source comes from molecules from which energy can be easily released. For example methane, present in the vented gases, taken up by bacteria (Zierenberg *et al.* 2000). There are no photosynthetic plants, or animals with vision, in these dark depths. The ecosystems comprise the whole gamut from bacteria to animals. Some of these organisms may have evolved around the vents and spread later to other sub-ocean vents. They may have later risen to the ocean surface. A few of the animals found there, such as blind frogs, most certainly arrived from above, but there has been speculation that life may have originated at these vents (Miller and Lazcano 2002). Some of the deep sea bacteria and fauna live close to the vents, at very high temperatures, others further away are exposed to the cold of the deep sea. The high pressure at these depths (which reaches 100 bar at a depth of 1 km) is not itself a great problem, as water is essentially incompressible.

Life below the land surface

An extremophile ecosystem which has recently raised much interest is the below-ground bacterial system (Gold 1999). As in the case of undersea vents, the bacteria derive their primary energy from energy-rich molecules formed in the heat from radioactivity still present in the depths of the Earth. Although the number of bacteria per unit volume is very small, the life-supporting volume, which extends down to a depth of about 3 km, is very great (below 3 km, temperatures become too hot for life). Gold calculates that the underground biomass may be larger than that of the total present in the oceans and on dry land. Moreover, contrary to common opinion, Gold believes that there are vast underground quantities of hydrocarbons (such as oil) whose origin is from chemical not biological activity. He believes that inorganically produced methane and other hydrocarbons, together with free hydrogen, may be the basic food of these bacteria. If Gold's ideas are correct, it may be necessary to completely revise our expectations of possible sites for extraterrestrial life, in addition to revising estimates of the world's oil reserves.

A note on extremophile mechanisms

Extremophiles are able to resist stress by numerous strategies. For example, if they grow in an environment that is rich in toxic salts, such as heavy metals, they may have mechanisms which either exclude or pump out the toxic ions. Heat-resistant forms have evolved with specialized proteins whose three-dimensional structure is stable at high temperatures at which normal proteins buckle and denature. Low-temperature organisms may have 'antifreeze' liquids in their bodies, which maintain liquidity at temperatures a few degrees below zero.

Halophytic monocellular *Dunalliela* algae grow in the almost salt-saturated water of the Dead Sea. They adjust their internal osmotic concentration to produce a water potential a little lower than that of the lake water (Ginzburg and Ginzburg 1985). Higher halophytic plants exhibit a whole range of salt resistance strategies. Among these are salt exclusion at the root level, salt excretion at the leaf level, various degrees of ion exclusion, and enzyme stabilization at the cellular level. They also have mechanisms for adjusting their cellular osmotic potential. For this, similar to Dunalliela, they often use innocuous sugars to balance external salts (Tartari and Forlani 2008). The osmotic adjustment is required to prevent desiccation in organisms growing in water liquid or to allow water uptake in land plants. The desert mice mentioned above have learnt to concentrate salt in their almost dry urine.

Many extremophiles live in their stressful environments not so much by adapting to stress as by avoiding it. Dormant bacteria and seeds were mentioned above. A further well-known example is the life cycle of ephemerals in hot dry deserts. These plants germinate, grow, flower, produce seeds, and die within a few weeks (usually in late winter) after enjoying a short period of rainfall. In this way they are not exposed to the difficult conditions prevailing throughout most of the year, during which they survive as dormant seeds.

Despite the great variety of extremophiles, which may or may not have evolved independently in their various niches, only one basic life form has been found on Earth (Chapter 6). All the extremophiles have the same basic cellular structure and their genes are built of the same four nucleotides as every other non-extreme organism on Earth. This is an important consideration when we try to calculate the chances of finding extraterrestrial life (Chapter 10).

Summary

- Evidence for evolution is found in fossils and other remnants left by organisms in soils and rocks, the isotopic remains of organic activity, and knowledge of the age of the strata in which the fossils are found. Strata are dated with radioactive isotopes.

- First life has been traced to the Archaean aeon, about 3.5–3.7 Gy BP. This is a relatively short time after the end of the period during which Earth was formed (the Hadean aeon) and the late heavy bombardment, during which life was not possible.

- From 3.5–0.5 Gy BP only unicellular life existed on Earth. Photosynthetic, single-celled plants evolved around 2.5 Gy BP. The first eukaryotic, multicellular life forms appeared around 545 My BP. Between 545 and 505 My BP, culminating at the latter date, there was a huge increase in multicellular life, the 'Cambrian explosion'.

- The first seed plants and insects appeared around 300 My BP. Reptiles and dinosaurs ruled between 200 and 65 My BP and birds and mammals from 65 My BP to the present.

- There were many ideas on the nature of evolution before the mid 19th century. They were best formulated by Wallace and then finally by Darwin. Darwin's thesis was based on 'natural selection' of a species, which appeared at random by mutation, that was most fitted to the environment and thus competed with other species. It has held ever since, despite many modifications, including its integration with modern genetics. Today it is recognized that evolution is not steady over time. There have been periods of fast and slow evolution ('punctuated equilibrium') affected by periodic environmental catastrophes (see Chapter 7).

- There are many current ideas on the mechanisms operating in evolution. Probably all are correct to some degree, or under some environmental circumstances. Among these are those borrowing from game and complexity theory and the theory of invasive symbiosis. The latter emphasizes cooperation rather than competition as driving evolution. Neutral and evolutionary development (Evo-Devo) theory emphasize the importance in evolution, in addition to mutations, of the rearrangement, activation, and suppression of existing genes, gene groups, and non-coding DNA present along the genome.

- Extremophiles are organisms which have developed on Earth in environmental niches unfavourable to other 'normal' species. Factors for which extreme tolerance has been developed in different species include: drought, heat, cold, salinity, acidity, alkalinity, ionizing radiation, and lack of solar radiation. Most extremophiles are found among the bacteria, but extreme stress tolerance characteristics are found in all levels types of organisms.

- Extremophiles are of interest as they extend the possibility of finding extraterrestrial life. There are a number of known environmental niches in the Solar System where unicellular Earth extremophiles could possibly live. Moreover, many characteristics of extremophiles (e.g. heat or salinity tolerance in plants) are of practical interest to breeders and 'molecular engineers'. Some extremophiles live under conditions which are presumed to have prevailed in the Archaean aeon and may shed light on the early origins of life.
- None of the extremophiles on Earth are very extreme in a cosmic sense. For example, none can grow without liquid water. Moreover, they have the same basic molecular structure (utilizing DNA and RNA) as all other Earth organisms.

9

The evolution of humans and their interaction with the biosphere

In this chapter the appearance of humans and their effect on planet Earth is reviewed. Maintaining the Earth's biosphere in a conformation favourable to humans is today an urgent priority. This urgency is the result of a rapidly increasing population and modern technology. There is a digression to a discussion of over-enthusiastic environmentalism. The latter, comes from genuine concern but is often based on poor information. It may engender policies which are misdirected. The wrong 'culprit' is often implicated. This can lead to a waste of always limited resources.

Human origins

Two hundred and fifty years ago the great Swedish taxonomist Carolus Linnaeus devised the system of plant and animal classification which has survived, with small changes, to this day. He assigned modern humans to the subspecies *Homo sapiens sapiens* (the very intelligent human; today Linnaeus might have called us *Homo horribilis horribilis*). The Linnaen genealogy for humans was as follows:

Kingdom, Animalia; phylum, Chordata; class, Mammalia; order, Primates; family, Hominidae; genus, *Homo*; species, *H. sapiens*; subspecies, *H. s. sapiens*.

The first Chordata (which later included the vertebrates) appeared around the time of the Cambrian explosion 550–530 My BP. The first mammals appeared around 220 My BP, but only expanded rapidly in the last 55 My.

Africa was the birthplace of numerous species of flora and fauna. The reason for this is not entirely clear, but is probably related to its position at the centre of the world's land masses. Moreover, climatic conditions suitable for life were better and more stable there than in more northerly regions. Central Africa was relatively free from the recurring ice ages of the Cenozoic era (Chapter 7). Consider a recent example: 6000 y BP, when much of Europe was still covered in ice down to the Alps, the Sahara region, now a desert, flourished, with forests, savannas, and accompanying wildlife.

The Hominidae appeared between 20 and 30 My BP and the gorillas and later the chimpanzees, from whom we are descended, 6–8 My BP. The genus *Homo* diverged from the Hominidae about 2 My BP. It formed a number of species, including *Homo erectus* and *Homo neanderthalis*. Remains of the latter have been found in the

Middle East and in southern Europe. *Homo sapiens* is thought to have evolved in sub-Saharan Africa 250,000–400,000 y BP. The exact timing of the appearance of modern humans, *H. s. sapiens*, is uncertain. Our nearest relatives, *H. neanderthalis*, survived up to a few thousand years ago. Before they went extinct, they coexisted with *H. s. sapiens*. Whether there was any interbreeding is a moot point.

For many years it was unclear as to whether *H. s. sapiens* evolved in Africa only or had multiple geographical origins. Recent analysis of the DNA of human mitochondria (mtDNA, which follows the maternal lineage) taken from numerous populations around the world indicates a common origin in Africa (Ingman *et al.* 2000). Best estimates put our ancestral 'Eve' as living in Africa about 200,000 y BP.

Analysis of mtDNA from extant African populations indicates that before humans moved out of Africa they lived in small isolated communities (see the URL link in References and Resources to this latest, ongoing, international research: data here are from initial research announcements in early 2008). Fossil and archaeological evidence indicates that humans began to spread from Africa to the rest of the world within the past 100,000 years and possibly as late as 60,000 y BP (Fig. 9.1). Note that humans only reached North then South America about 20,000 y BP, when the Bering Strait land and ice bridge still connected Asia and North America.

There are a number of important characteristics common to humans and other Hominidae; these include an upright gait. This allowed the hands, which have opposite

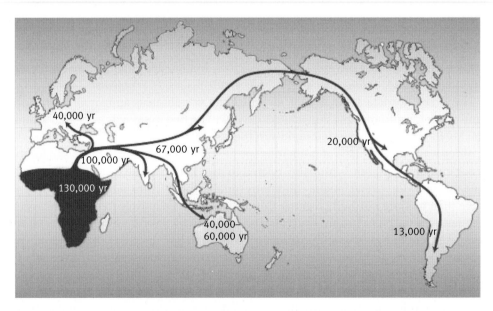

Fig. 9.1 Dispersal of modern humans from Africa to the rest of world. The dates are from findings of the earliest presence of *Homo sapiens sapiens* in the areas indicated. Courtesy of Nature.

grasping thumbs and fingers, to be used for making tools. The most important development of *H. s. sapiens* is of course brain function. Humans developed the ability to communicate with each other: speech, writing, books, and today the electronic media. Other characteristics of our brain such as introspection and emotion are often toted as unique, but may also be shared by other animals such as dolphins, monkeys, elephants, and dogs.

Brain size is certainly important. The brain of *H. s. sapiens* has a volume of about 1350 cm^3, more than double that of a chimpanzee or gorilla. However, elephants have brains some three times larger than that of humans. Sperm whale brains are larger than ours by a factor of four. Cranial volume or brain weight is not the sole criterion for mental capacity. For example, at the lower end of the scale, bees carry out advanced functions requiring both built in (pre-programmed) and responsive thinking. This includes vision and tactile sensing of the environment. The information they gather is analysed and used for flying, fighting, and foraging for food. They communicate with other bees and transfer intelligence on food sources, navigation and fighting. They also maintain homes (villages) with a very strict social organization, which includes building functional architectural structures, allocation of duties, reproduction, defense, food storage and emergency contingency planning. All of this activity is coordinated with a brain the size of a sesame seed (Haddad *et al.* 2004). Brain complexity is certainly involved in bees and in us, but our understanding of brain function is rudimentary.

Homo sapiens sapiens is often considered to be the species at the very pinnacle of evolution, but our emergence was fortuitous. Our position on the evolutionary tree (Fig. 8.1) is anthropocentric. We of course drew the tree; a dolphin may have drawn Fig. 8.1 with itself at the top! As noted, humans appeared only very recently. We might have emerged before, within any animal group having a well-developed nervous system. Why not an intelligent dinosaur? Contrariwise, we might not have appeared at all. Although life has been present on Earth for the last 3.7 Gy (a quarter of the entire lifetime of the universe) and animals with advanced nervous systems have existed for hundreds of millions of years, *H. s. sapiens* appeared only once. It required only a minor change in our genes to differentiate us from the 'lower' species. Our genome differs by only some 2% from that of chimpanzees and 0.5% from that of Neanderthals. However, although these small numbers are humbling, recent advances in our understanding of gene function indicate that they may be hiding a more elaborate reality. Evolutionary advances are made not only by the selection of new gene mutations, but also by the rearrangement, activation, and inhibition of existing genes and intergenic material (Chapter 8).

As discussed further in Chapter 10, from an evolutionary standpoint it has yet to be proven that *H. s. sapiens* is a successful new species. There are two main criteria for evolutionary success: fecundity and survival. Humans have certainly shown that they are fertile and have spread to all parts of the globe. Survival is another matter. Our increase in population and technological abilities are counterbalanced by our poor social talents. This is expressed in our inability to overcome regional population explosions, hunger, pollution, disease, wars, and depletion of resources. Solutions exist today for most of these problems, only local and international determination is lacking. As discussed below, much of this conflict with the environment (and each other) is caused by our tendency to overuse and pollute natural resources. Add to this the inability to overcome the worst of our inherited animal instincts, especially the 'territorial imperative'. The average longevity (from appearance to extinction) of advanced animal species has been of the order of 1 million years. It is uncertain whether humans, who have been around for at most 200,000 years, will survive another millennium.

Human population

Its not that people started breeding like rabbits, it's just that they stopped dying like flies.

UN consultant Peter Adamson, quoted in Lomborg (2001)

The first millennia of humanity (*H. s. sapiens.*) left us no census data and the numbers are hard to reconstruct. Best estimates put the original African population in the tens of thousands at most. 70 y BP the Indonesian volcano Toba erupted in a devastating cataclysm whose world-wide effects reduced all life on Earth. It is estimated that at that time the entire world population of humans was reduced to no more than some 15,000 individuals. The population remained very small up to the late Stone Age (Ambrose 1998). This was the period between 7500 and 3800 y BP, during which the first polished stone tools were in evidence. One of the most thorough studies of early human populations can be found in Hawks *et al.* (2000).

The last ice age ended around 11,000 years ago. At this time humans began to flourish, probably as a result of the more favourable climate extending the food-gathering areas. This was certainly the case in southern Europe, North Africa, and the Fertile Crescent of the Middle East, from which we have most records. Humans went from living in family groups of hunter gatherers to farming. Villages and towns soon followed. Sedentary life encouraged industry, trade, traders, and culture. Early culture, especially writing, mathematics, astronomy, and medicine, developed at this time (after about 4000 BCE). This happened in Egypt, Babylonia, Assyria, and elsewhere in Asia and the Far East. On an evolutionary time scale these developments were concurrent.

Even with the development of urban civilization, human populations remained quite small for thousands of years. As shown in Fig. 9.2, the total human population is estimated to have remained low, rising slowly to the range of millions in the Middle Ages. In the centuries around 1000 CE the human population was reduced, at least in Europe, by the ravages of disease. This was associated with a long period of higher temperatures, which were also detrimental to agricultural crops. Such events almost certainly had occurred before and in other regions of the world, but went unrecorded.

As Fig. 9.2 shows, world human population increased exponentially after the Middle Ages and reached its first billion at the beginning of the 19th century. Estimates of future world population size are of course speculative, but even the most cautious prediction indicates an increase from 6 G at the beginning of the 21st century to 8 G in around 2050,

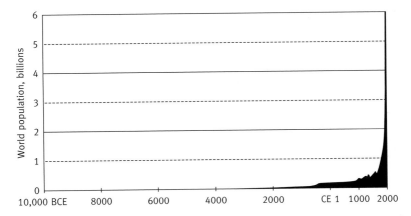

Fig. 9.2 Estimates of the human population in the last ten millennia. (Redrawn from various publications of the UN)

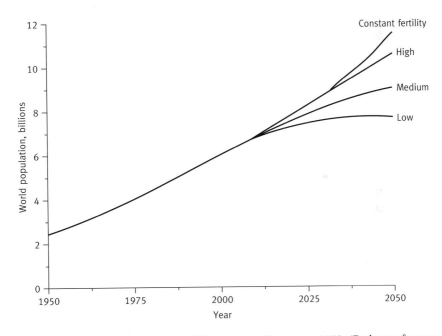

Fig. 9.3 World human population from 1950, with predictions to 2050. (Redrawn from various publications of the UN)

with a mean estimate of 9.83 G (Fig. 9.3, from data of the UN). Recent, encouraging, world population data show that even in the developing nations the birth rate has been slowing down. The 7.92 G estimate for around 2050 takes this into account. Demographers do not agree on the reason for the downtrend in fertility; there are many possibilities. There does seem to be an inverse correlation between the education of mothers and the number of children born.

The predicted population shown in Fig. 9.3 embodies many highly significant demographic changes, in addition to the 50% increase in numbers. Table 9.1 gives the annual population growth in a number of selected countries as published by the UN in 2007 (based upon data and estimates for 2005–10). As seen, some of the most populous countries, and the poorest among them, are those with the most rapidly increasing populations. For example, the population of Bangladesh, 159 million, is increasing by 1.67%

Table 9.1 Population and annual population growth rates for a selection of countries

Country	Present population (millions)	Annual growth (%)
Afghanistan	27.1	3.85
Algeria	33.9	1.51
Argentina	39.5	1.0
Bangladesh	158.7	1.67
Belarus	9.7	−0.55
Brazil	191.8	1.26
Burundi	8.5	3.9
China	1328.6	0.58
Congo (Democratic Rep.)	62.6	3.22
Egypt	75.5	1.76
Estonia	1.3	−0.35
Ethiopia	83.1	2.51
France	61.7	0.49
Germany	82.6	−0.07
India	1,169.0	1.46
Indonesia	231.6	1.16
Italy	58.9	0.13
Japan	128.0	−0.02
Mexico	106.5	1.12
Nigeria	148.1	2.27
Pakistan	163.9	1.84
Philippines	88.0	1.9
Poland	38.0	−0.15
Romania	21.4	−0.45
Russian Federation	142.5	−0.51
South Africa	48.6	0.55
Thailand	63.9	0.66
Turkey	74.9	1.26
Uganda	30.9	3.24
United Kingdom	60.8	0.42
United States	305.8	0.97
Yemen	22.4	2.97

Data as published by the UN Department of Economics and Social Affairs (2007), and based upon data and estimates for 2005–10.

(2.66 million) every year. This annual increase is twice the total population of Estonia, which is decreasing.

The increase in world population is not uniform. Geographical and age-wise distributional changes will result in enormous social pressures (Cohen 2005). Europe differs from the general world trend, and is currently suffering a decrease in population. The world's most populous country, China, has slowed its rate of population increase by political decree. Its population of 1.35 G may soon be overcome by that of India. India's population is currently 1.17 G (2007 UN figure) and is increasing by 1.46% annually (an extra 17 M individuals a year). The lowered birth rate in the more developed countries results in an ageing of the population. This puts considerable pressure on the workforce. Moreover, there are not enough young people entering the workforce and making payments to the pension funds, which are often unable to cater for the larger (percentage wise) retired sector. It is predicted that, with its negative birth rate (Table 9.1), half the population of Japan will be older than 50 by 2050. China is already showing these demographic stresses as a result of the planned reduction in birth rate. In the less developed countries where most of the population increase is taking place, the situation is reversed. Young people make up the majority of the population and they are having more and more difficulty in finding profitable employment.

The spread of modern agriculture exacerbates the employment problem. At the beginning of the 20th century, 60% of the population in the more developed countries was engaged in agriculture. Today the figure is 2–3%. At the same time agricultural production and food per consumer, have increased greatly. This has brought about mass migration from the villages to the cities. As agriculture becomes more advanced in the developing countries the same trend is evident, with the addition of migration to the more developed countries.

Before the Second World War the populations of the European countries were relatively homogeneous. Today they are receiving influxes of immigrants from Africa and less fortunate Middle Eastern and Asian countries. For example, France is now home to some 6 M immigrants from North Africa. This is 10% of the population, with the percentage increasing rapidly as the new immigrants are the most fecund section of the population. Germany has received about 2 M immigrants from Turkey; the UK has numerous minority groups which came from Pakistan, India, and other former colonies and, in recent years, from less wealthy eastern Europe.

Contrary to most other developed countries, the USA is currently showing an increase in population (Table 9.1). This is not so much a result of the birth rate of the original population, as of immigration, which of course has been building the USA since its inception. The original immigrants came mostly from Europe, with the later addition of slave people from Africa and semi-slave workers from Asia. An example of the change in population composition can be seen in California, the most populous state of the union. In the last 50 years the population (92,000 in 1850, 37 M today) has gone from mainly European Caucasian, to a majority of people of Latin American, Asian, and African origin.

Homo sapiens sapiens (*horribilis horribilis*?) and the biosphere

The disastrous effect of plants on the biosphere (i.e. from the point of view of plants!) was described in Chapter 6. Plants, it will be recalled, used up a large amount of their required resources, especially CO_2. The oxygen they emitted as waste poisoned the environment, both for themselves and for many anaerobic organisms such as the essential nitrogen-fixing bacteria. The high oxygen atmosphere they produced was the very environment required for the evolution of plants' natural enemies. The effect of humans on the biosphere may be even worse. This is not only from the point of view of humans but also of that of many other species which are being driven to extinction. There are innumerable books and articles on the negative environmental effects of humans on the biosphere. Only a few of the main points can be briefly mentioned here.

Environmentalists tend to be prophets of doom. We humans seem to enjoy scary scenarios; they make great movies. Unfortunately such scenarios are often right. Even when exaggerating, blaming the wrong cause, or suggesting an incorrect solution, they may draw our attention to an important problem (see for example the best-seller by Diamond, 2004). Lomborg (2001) has presented a detailed study debunking many excesses of 'environmental politics'. It attracted the expected outrage of offended ecologists, but is worthy of serious consideration.

Some of the main damaging effects of humans on the biosphere

Air pollution

Humans pollute the atmosphere with both gas and particulate effluents (Jacobson 2002). The most recognized gas pollutant is CO_2, and its increase is indeed very dramatic. As shown in Fig. 9.4, the CO_2 in the atmosphere is now some 38% higher than before the 19th century and the onset of the Industrial Revolution. The original CO_2 component of the atmosphere came from volcanoes (Chapter 5), but today volcanoes are responsible for only some 1% of the added CO_2. The burning of fossil fuels produces most of the increasing CO_2 content—about 27 G tonnes of CO_2 annually. Deforestation and burning are the second major source. According to data from the United Nations Environmental Programme (UNEP) the world loses about 16 M hectares of forest annually. Not all the CO_2 released remains in the atmosphere: some is dissolved in the oceans and fresh waters and, as discussed below, some is taken up by plants, increasing the standing mass of the world's vegetation (see below).

The increase in atmospheric CO_2 is commonly considered to be bad for humanity, as it increases the greenhouse effect (Chapter 3) and raises world temperatures. The quantitative reliability of this conclusion is questionable. It is discussed below.

Many of the other gases and airborne particles released as a result of human activities, such as lead derivatives, have very harmful effects on animals (including

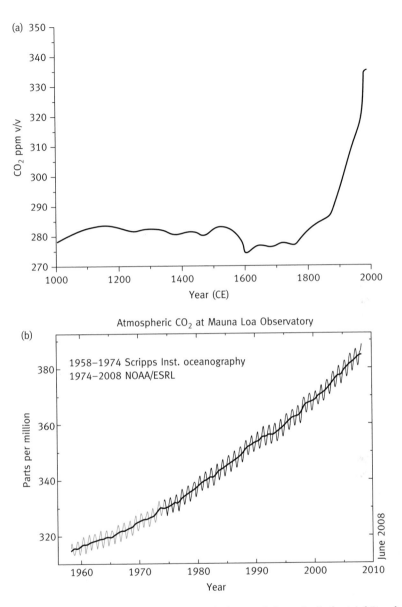

Fig. 9.4 The CO_2 concentration in the atmosphere before and since the Industrial Revolution: (a) since 1000 CE; (b) since 1960. Data were collated from a number of sources: before 1850, mainly from air trapped in ice cores; in recent years from direct atmospheric measurements.

humans) and on plants (Omasa *et al.* 2006). Gases other than CO_2 also contribute to the greenhouse effect, especially methane and nitrous oxides.

Methane (CH_4) is still present in the atmosphere in relatively low concentrations, but its absorptivity of near-infrared radiation (which is the basis of the greenhouse effect) is many times greater than that of CO_2.[1] Methane is released in the flatulence of cattle and sheep. Their numbers are growing, together with the world's human population and an increase in meat consumption per person. Another source of methane, as large as or larger than that from cattle, is fields of paddy rice. Microbes in the wet paddy mud release vast quantities of methane. On a global scale this is estimated to be 100 M tonnes a year. There is a recent report indicating that under anaerobic conditions living plants may also release large quantities of CH_4 (Keppler *et al.* 2006). Another potential source of CH_4 is from the permafrost of the Arctic (see below in relation to climate change).

The water vapour content of the atmosphere increases with water temperature, and varies greatly in different regions. Water vapour is a major contributor to the greenhouse effect. It exerts a feedforward action: its concentration, and therefore contribution to the greenhouse effect, increases as the world's temperature increases.

Human activity releases vast quantities of dust and aerosols. These tend to increase the atmospheric albedo, reflecting back the incident solar radiation. On the other hand, dust increases the solar radiation absorptivity of the atmosphere and, under some conditions, catalyses water nucleation, increasing cloud formation. The overall effect of dust and aerosols on the world's heat balance has proven difficult to quantify. This is a major source of uncertainty and error in the mathematical simulation models used for short-term weather forecasting and for analysing and predicting the long term effect of greenhouse gases.

The UV radiation penetrating the atmosphere interacts with many industrial pollutants to produce ozone (O_3). Ozone in the air near the surface of the Earth is toxic to both animals (including humans) and plants. Its effects are often insidious and difficult to detect and quantify at low levels. Contrariwise, when ozone is present in the stratosphere[2] it is a positive factor. It is formed there from the action of UV solar radiation on oxygen, but is itself a much more powerful absorber of UV than O_2. Life could not exist on the surface of Earth were it not for this UV attenuation, especially in the UVB waveband (270–315 nm).

Chlorofluorocarbon gases (CFCs) and the bromofluorocarbon gas halon are used in industry and CFCs are used domestically in refrigerators. When these gases reach the upper atmosphere they are broken down by the action of sunlight (photodissociation). This releases atomic chlorine and bromine, which catalyse the breakdown of O_3 back to O_2. In the 1970s and 1980s a gradual drop in stratospheric ozone, of about 4% per decade, was recorded. The effect was strongest in an area over the Antarctic, the so-called 'ozone hole'. This depletion resulted in an increase in the amount of UV radiation reaching the

---

[1] Unlike CO_2, methane slowly breaks down in the atmosphere. Even so, over a 20 year period its contribution to absorbance of near-infrared radiation is 25 times that of CO_2, on a mass per mass comparison.

[2] The layer of the atmosphere extending from about 30 to 50 km above the Earth's surface.

biosphere. One of the few victories for environmental legislation on a world-wide scale was the banning of the use of CFCs (by the Montreal Protocol of 1987).

Compliance was very encouraging. Among other things, CFCs are no longer being used for new refrigerators. In the years that followed the ozone hole appeared to have become smaller. Unfortunately, recent figures indicate that it may be growing again. Moreover, Son *et al.* (2008) suggest that a weakened ozone layer may have other, perhaps negative, effects on Antarctic weather patterns. Their conclusions are based on new computer models. This is an example of the great difficulty in predicting future world weather and is very relevant to the discussion of human effects on climate change (see below).

Another anthropogenic gas which is playing havoc with many ecosystems is sulphur dioxide (SO_2). It is released when sulphur-rich coal and liquid fuels are burnt. Burning sulphur-rich coal and oil in power plants (which goes back to the beginning of the Industrial Revolution) is becoming more and more widespread. This is because of the high cost of low-sulphur coal and oil and the lack of economically competitive alternative energy sources.[3] Apart from the direct toxic effects of SO_2 on plants and animals it, together with acidic nitrogen compounds released into the atmosphere, produces acid rain when washed out by precipitation. Acid rain reduces the growth of forests and agricultural crops, affects the flora and fauna of lakes, and can very adversely affect the microflora of soils (Seinfeld *et al.* 1998).

Water use and pollution

We live on a planet having 71% of its surface covered with oceans (Chapter 1), but humans are creatures of dry land and require fresh not sea water. Fresh water makes up only 3% of Earth's total water, and 77% of that is sequestered in the polar ice caps. We use relatively vast quantities of fresh water for agricultural, industrial, and domestic requirements. At the same time we pollute many of our water sources, making them unusable. It is consequently of little surprise that with the burgeoning human population the availability of fresh water is becoming a major problem (Rennie, 2005).

The following are just a few brief notes on the world water crisis, both ocean and fresh, emphasizing some topics worth special attention:

• Although the seas are vast they are affected by human effluents. In 1947 Thor Heyerdahl made his fabled crossing of the Pacific Ocean on a raft, which kept him in close proximity to the water. One of his observations was that in the centre of the ocean, thousands of miles from land, the waters were polluted by breakdown-resistant plastic detritus floating on the surface. More than 60 years have now passed and the situation is only

---

[3] The mid 2008 price of oil (to above $140 per barrel) could have changed the economics of many alternative sources of energy, such as photovoltaics, wind, geothermal, tidal, etc. (but in early 2009 its price fell to $40. See Appendix B).

worse. There are two major factors to be considered. The first is that a large proportion of Earth's human inhabitants live close to the sea. The second is that human excrement and other domestic and industrial effluents and wastes, such as Heyerdahl's floating plastics, are frequently released into the sea without any prior treatment.

- Most of the fisheries of the world are showing a very considerable (20–90%) drop in catch. This is affecting many people living in coastal areas who derive much of their dietary protein and income from the sea. On a larger scale, by way of example, the West Atlantic adult bluefin tuna population is estimated to have gone from 300,000 individuals in 1970 to 25,000 in 1990. However, this decline has not been caused by pollution. As in many other regions of the oceans, it is a result of intense fishing and not a reduction in productivity. Populations of those fish blessed by having no commercial value have not shown a decline in recent years. When fishing stops fish stocks tend to rebound.

- Shortage of fresh water is endemic in all the arid and semi-arid regions of the world, especially in large parts of Africa. There are also shortages in less arid, advanced countries, especially in large urban areas such as London. The UNEP calculates that 300 million people in Africa are currently suffering from a chronic shortage of fresh water. Not all water shortages are a result of over-use by humans, but they are often exacerbated as numbers increase and living standards rise. The classic example is Lake Chad, in central Africa, which supplies fresh water to four countries. It has historically expanded and contracted with changing rainfall patterns. Today water use by the increasing human population, exacerbated by an extended period of low precipitation, have so reduced the lake that it may disappear entirely in the next decades, putting tremendous pressure on the local peoples.

- By far the biggest consumer of fresh water is agriculture which, in semi-arid regions of both developed and developing countries uses about 80% of what is available. Domestic use of water varies enormously, from about 300 litres per person per day in developed countries to 6 litres per person per day in much of Africa. Domestic, industrial, and commercial use account for the remaining 20%. The exact percentages used by the different sectors vary greatly from country to country, as does the percentage of the potentially available water being put to use.

- From the late 1960s India (and some other Asian countries such as the Philippines and a little later China) underwent a 'Green Revolution'. The introduction of high-yielding grains, irrigation, and fertilizers changed India from a food-importing to a food-exporting nation. Formerly, it had suffered periodic famines. The greatest agricultural development was in the Punjab, which has much land but inadequate rainfall. Since the turn of the century, food production has levelled off. With the rising population this means less food per person, and food is again being imported. One of the main reasons for this is that fresh irrigation water has become a limiting factor. Underground aquifers are being exhausted, as shown by the drop in the level of water tables. Moreover, government development of canals, bringing water from rivers to the semi-arid grain districts, lags far behind requirements.

- The world as a whole is not short of fresh water. The problem is geography, the cost of distribution, and pollution. Canada, for example, has about 2 M fresh water lakes, while more southerly California often suffers from severe droughts but does not import water from Canada. Pollution of fresh water sources is so rampant that even in the USA, where there is an awareness of environmental problems, about 40% of all lakes cannot safely be used for swimming.
- An 'obvious' solution to the shortage of fresh water is desalination of the almost unlimited ocean waters. Whether to do so is a question of cost. Even with modern reverse osmosis systems, much energy is required to pump water through the salt-selective membranes. The Emirates of the Persian Gulf occupy a region desperately short of fresh water. They operate more desalination facilities than any other part of the world. There, cheap oil is essentially being converted into expensive fresh water. This is not an option for most countries.
- In many countries of North Africa, the Middle East, and elsewhere 'fossil' water, left underground since the ice ages, is being tapped. It often supplies much of the fresh water needs; e.g. in Libya 40% of the fresh water comes from such a source. These aquifers are limited and in many places will soon be exhausted.
- Another 'obvious' way to extend fresh water supplies is to prevent waste and recycle whatever is available. For example, it has been calculated that about half of the domestic water distributed in London, Mexico City, and other metropolises is lost in leaky, centuries-old delivery systems. Their repair requires huge capital investment. In advanced agricultural regions, such as California and Israel, irrigation water is being conserved by drip irrigation technology. Moreover, in these countries much of the domestic water used is collected and purified to a level acceptable for agriculture. In the developing regions there is much room for greater use of these technologies.
- The world wide shortage of unpolluted fresh water remains. In many regions, especially the more arid, the increasing population puts tremendous pressure on water supplies. This can quickly become a *casus belli*.

Radiation pollution

It is difficult to make only a brief mention of a pollution source which, today or tomorrow, could send most or all of us into oblivion. This could happen if there was a war in which nuclear weapons were used. On the other hand, in a more optimistic vein, it is possible to argue that fear of mutual destruction by radiation was the factor responsible for peace during the 40 years of the USA–Soviet Cold War. This was possible given a modicum of humanity and common sense, which seemed to have prevailed after the horror of nuclear war was demonstrated on the unfortunate Japanese toward the end of Second World War.

At the beginning of the 20th century, Marie and Pierre Curie laid the foundations of our understanding of radioactive materials. They discovered radium and polonium (for which, and for other work, they received the Nobel prize). Unfortunately they had little understanding of the health hazard of exposure to the ionizing radiation emanating from these unstable compounds. Marie Curie died in 1934 of aplastic anaemia, probably caused by her long exposure to radioactive elements.

Atomic weapons remain the major potential source of dangerous environmental radiation. Up to a short while ago they were tested indiscriminately by the nuclear powers, with little consideration for the world-wide increase in radioactive contamination of the atmosphere and oceans. Nuclear power plants have many advantages but are prone to terror attacks, human error, and mechanical breakdowns. The worst such accident, so far, was the explosion at the Chernobyl nuclear power station in the Soviet Union (Ukraine) in 1986. The fallout and cloud of radioactive material which ensued were calculated to have been more than 20 times that resulting from the atomic bombs dropped on Hiroshima and Nagasaki. Dozens died immediately or within weeks and the long-term damage to the nearby (and not so near) population was enormous, but cannot be exactly assessed. Another major problem with nuclear power stations is disposal of radioactive waste which may stay active for hundreds of years. So far no really satisfactory solution has been found other than sequestration deep underground. Even so, given the shortage of energy from environmentally friendly sources, and the increasing cost of oil and coal, many believe that increasing reliance on nuclear fission energy may eventually be inevitable. Theoretically, hydrogen fusion, as occurs in the Sun (and hydrogen bombs) is the most attractive nuclear energy source. It requires no uranium or other nuclear fuel (just water) and gives off little waste. Unfortunately, decades of research have failed to solve engineering problems of containment. Safety and security problems would of course remain.

Today we are exposed to radiation from a number of different sources in our daily lives. These include X-rays and medical radioisotopes, leaky microwave ovens in our kitchens, and even natural radon gas. The latter comes from the Earth and may accumulate in poorly ventilated cellars or buildings in some geological regions.

Not all radioactivity is without value to humans. Radiation and radioactive isotopes are used in medicine and biological research, and the extraordinary importance of the latter in geological and archaeological dating was discussed above. A comprehensive collection of articles on many aspects of 'man-made'[4] and naturally occurring radioactivity can be found in Tykva and Berg (2004).

Homo sapiens sapiens and species extinction

Make no mistake, extinctions are a normal part of evolution. Not counting unicellular organisms, some 99% of all species which appeared on Earth have become extinct. In Chapter 7, catastrophic events which afflicted life during its 3.7 Gy history were shown to have periodically caused mass species extinctions. But even without these events the

---

[4] Quoting the authors: a clear if 'politically incorrect' title.

average lifetime of a species has been estimated to be between 100,000 and 1 million years (Lawton and May 1995). This is only a small fraction of the 545 My of the Phanerozoic aeon.

Today, many biologists believe that we are in the middle of a new, human-caused Holocene extinction event, which became most noticeable after 1500 CE and is now accelerating rapidly. The rate of extinctions appears to have increased by a factor of between 100 and 1000. It has been estimated that 20% of all extant species will become extinct within 30 years. Wilson (2002) believes that, if the present trends continue, 50% of all species will disappear by the end of the 21st century. One example, out of many sad instances, is the dodo (*Raphus cucullatus*). It is one of the most fabled animals to have become extinct in recent historical times. This flightless bird was first found by European explorers when they reached the island of Mauritius in the late 16th century. It appears in many illustrations of that time (Fig. 9.5). It was hunted for food until the last of its kind was reported in 1681. (Some ecologists believe that the extinction numbers are exaggerated, with too much emphasis being put on advanced species (Gibbs, 2004)).

The reasons for the accelerated extinction of species, caused by humans are quite clear:

- introduction of alien species which compete with the endemic flora and fauna;
- loss of natural habitats; e.g. the bamboo forests of China, long the foraging grounds of the Giant Panda (*Ailuropoda melanoleuca*) which eat only bamboo leaves;
- over-exploitation, such as the now extinct dodo or the West Atlantic tuna, mentioned above.

(See the list of URLs in References and Resources for two portals to the entire field of human-induced extinctions.)

Politically correct environmental science

People with an interest in science but who do not themselves practise this calling tend to set scientists on a virtual pedestal of integrity. Their findings are thought to be completely objective, based on solid, unchallengeable facts. Scientists know this to be otherwise. Apparent 'facts' tend to change with each advance in science. Moreover, scientists are human. They often set out to prove a hypothesis, instead of to test it. Environmental science is rampant with insufficiently supported conjectures (Lomborg 2001).

Because of the often partly correct but emotionally loaded environmental causes, some of them become politicized. There is danger in exaggerating environmental problems and their proposed solutions. The 'cause of the week' may end up deflecting attention from the real culprit. Repair of environmental problems is usually expensive and misidentification of their cause may result in wasted resources. The following are four examples of sometimes incorrect and sometimes misdirected, if well intentioned, environmentalism.

Fig. 9.5 A model of the dodo (*Raphus cucullatus*), last seen on Mauritius in 1681.

The Amazon rain forest

It is often stated that we must stop cutting down the Amazon rain forests because they are the 'lungs of the world', taking up CO_2 in photosynthesis and emitting oxygen. There are many excellent reasons to stop logging in the Amazon, but this is not one of them. In a mature ecosystem such as this there is no *net* growth in terms of additional biomass accumulated per unit land area (ecologists call this the 'climax' of the ecosystem). There is annual growth, with CO_2 being taken up and oxygen released to the atmosphere, but it is balanced by the decomposition of the seasonally discarded dead leaves, branches, and fruit. This detritus falls to the forest floor where it is consumed by the soil microflora. In the course of their respiration, the microflora take up oxygen and release CO_2 to the atmosphere in the same quantities as pass in the reverse direction during photosynthesis. Only when the forest is burned is there a net output of CO_2 into the atmosphere. It is the continuing forest clearing and burning in the Amazon region which is a major contributor to the rise in atmospheric CO_2. At the same time as the CO_2 is released in fires almost the same quantity of oxygen is consumed, but it has little percentage effect on the huge concentration of oxygen in the atmosphere (21% or 21 kPa). This is contrary to the effect of CO_2, which is added to an atmosphere with only 0.038% (38 Pa) CO_2.

Increasing atmospheric CO_2

The anthropogenic rise in the CO_2 content of the atmosphere (Fig. 9.4) is said to damage the entire biosphere, but we should recognize that it also has advantages. Vegetation, a major component of the biosphere, may benefit from the increase in CO_2.

It was noted in Chapter 6 that, as a result of the world-wide success of photosynthesizing plants, which evolved about 2.6 Gy BP, the CO_2 content of the atmosphere was eventually reduced to below what is optimal for plant growth. This became especially critical and continuous in the Cenozoic era, which began 65 My BP and extends to the present. During this era, atmospheric CO_2 fell to ~26 Pa (260 ppm v/v). This was almost certainly a major evolutionary driver for the emergence of C_4 plant photosynthesis[5] from the original C_3 mechanism. C_4 metabolism is more efficient in acquiring CO_2 from an atmosphere with a low CO_2 concentration and is more resistant to high (21 kPa) oxygen levels (von Caemmerer and Furbank 2003). Despite these advantages, but for a number of other reasons (mainly the requirement for a higher light intensity), only 18% of the world's vegetation belongs to the C_4 group.

It follows that the very recent (last 150 years) anthropogenic rise in atmospheric CO_2 (from 25 to 38 Pa) is often beneficial to the Earth's C_3 type plants (82% of species). Wherever root growth, light, water, and mineral nutrition are not limiting, the photosynthesis and growth of C_3 plants is found to increase in response to today's elevated atmospheric CO_2 and certainly to the 60 Pa level predicted for the coming decades (Long *et al.* 2004; Tuba 2005). Moreover, as Long and others have shown, there are secondary positive effects on vegetation of increased CO_2, such as lower water use per unit dry weight acquired and greater resistance to heat and soil salinity. This can be expected to increase plant growth on the edges of dry deserts and move the vegetation front in the direction of increasing aridity and salinity. This effect will be more pronounced as the CO_2 rises from 38 to the predicted 60 Pa. It is also expected to increase food production per unit area of land and volume of water (Wittwer 1995).

A rise in global photosynthesis in response to the higher levels of CO_2 has been deduced from the curve of increasing atmospheric CO_2. As seen in Fig. 9.4(b), in the Northern Hemisphere there is an annual drop in the level of CO_2 during the summer

---

[5] The timing of the first appearance of C_4 metabolism is uncertain. It is estimated to have evolved around 150 My BP.

months. Idso (1986) suggested that the amplitude of this reduction was increasing with time, as a result of world-wide photosynthesis responding to the rising atmospheric CO_2. This, of course, is just a correlation. The annual fluctuation in the curve of rising CO_2 was also analysed by Keeling *et al.* (1996) and by Myneni *et al.* (1997). They suggest that the increased photosynthesis may be a result of the concurrent warmer weather and longer growing seasons, which they relate to climate change in the wake of rising CO_2.

The fact that rising CO_2 may benefit vegetation (including agricultural crops) does not mean that it is generally a good thing! Burning of forests and fossil fuels, which releases this CO_2, is obviously detrimental. It also contributes to global warming, although the degree to which it does so is disputed (see below).

Biofuels

It has been said that pollution from CO_2 and nitrogen derivatives in car exhausts and the shortage of fossil fuel can both be offset by using ethanol. According to proponents of ethanol its combustion produces 20% less CO_2 than fossil fuels and no nitrogen gas pollution. It can be produced from vegetation which, in the course of its growth, takes up CO_2 from the atmosphere, thus reducing the CO_2 greenhouse effect. This idea has been widely touted by environmentalists. Conversely, mid-2008 production of biofuel was blamed for causing a sudden world-wide food shortage and consequent increase in food prices. Both claims were highly exaggerated, if not plain wrong.

As in most such arguments reliable data are often lacking. Growing plants basically convert solar energy into plant material. At best they do so with an overall efficiency of about 4%. To achieve even such a low yield requires good soil, sun, water, fertilizers, oil-driven tractors, etc. When the energy costs of growing crops and converting them to ethanol are added up, including: the production of fertilizers and manufacture of farm machinery, harvesting and production of ethanol, and the cost of distribution, the energy balance is often negative. Moreover, CO_2 pollution is usually increased.

Brazil has been the pioneer and leader in ethanol technology. But there conditions are rather special. A first factor is that Brazil has large regions of land suitable for sugar cane production. Sugar cane is a C_4 plant (see above) with very high yields. Second, Brazil must spend its scarce foreign currency to purchase the fossil fuel required for the vehicles driven by the 20% of the population who own cars, and for the industrial sector. Ethanol production raises the cost of food, but that hardly affects the high-income sector. In Brazil, land for growing sugar cane is becoming scarcer. The problem is being addressed by cutting down and burning swaths of pristine Cerrado savannah and Amazon forest. This results in huge quantities of CO_2 being released into the atmosphere. The situation is similar in Malaysia and Indonesia. In these countries large areas of land are occupied by oil palm trees. The oil produced is an essential part of the population's food and cooking fuel. In recent years it has been sold for the production of diesel fuel. This has increased the price of palm oil, severely affecting the poorest part of the population. At the same time forests are being cleared, burnt, and replanted with more oil palms. This has a long-term negative effect on the natural eco-systems of these tropical countries.

In the last few years the concept of ethanol replacing gasoline was introduced into the USA, some European countries, and elsewhere. Farmers have been subsidized to produce crops for ethanol production, resulting in a large move in the USA to prod-uce corn (maize) for ethanol production. The Pulitzer prize-winner Thomas Friedman wrote an enthusiastic endorsement of ethanol production in the 20 September 2006 edition of the *New York Times*. Many others joined the bandwagon.

For some years reports in the science literature have indicated that, at best, only some crops growing on poor lands, such as switchgrass, may show an environmental and eco-nomic advantage when converted to ethanol (Schmer *et al.* 2008). A more attractive bio-fuel is biodiesel produced from the conversion of agricultural waste. This technology would also reduce other pollutants (Hill *et al.* 2006), but waste conversion still has severe production problems, especially the breakdown of cellulose. It is potentially only a mar-ginal source of energy. Searchinger *et al.* (2008) calculated the CO_2 emission cost of the conversion of land to corn production for ethanol. They found that, in the long run, the increased release of CO_2 far exceeded the savings in emissions. A similar negative envir-onmental impact was the conclusion of Scharlemann and Laurance (2008) who reviewed a detailed, 2007 Swiss study, on the CO_2 balance of many alternative fuel crops. Despite the generally negative results of these studies, countries lacking foreign currency for fossil fuel such as Brazil, may still be attracted to local biofuel production, despite the impact on the cost of food.

The enthusiasm for biofuels suddenly reversed at the beginning of 2008. At that time world food prices shot up sharply as supplies dwindled. Shortage of basic food staples on a world scale is a market tendency which started around 2004. The 'politically cor-rect' view in mid-2008 was that the entire ethanol technology was a scam and is respon-sible for the present world-wide food shortages (e.g. Paul Krugman, 'Biofuels...a terrible mistake', *International Herald Tribune*, 8 April 2008). A look at the numbers shows that this too is an exaggeration. While corn prices in the USA and cooking oil in Malaysia and Indonesia have certainly increased in price as a result of their being sold for biofuels, they represent only a small percentage of overall world food production.

As Krugman pointed out, there are a number of other major reasons for the world food crisis. Crop yields in the last years, prior to 2008, have been relatively low, espe-cially in Africa and Australia. The decreasing food supply in India was discussed above, in relation to the dwindling availability of irrigation water. Of even greater importance is the increasing numbers and greater purchasing power of the populations of the great developing nations in Asia, foremost among them China and India. Instead of suffer-ing recurring famines, they now have the means to purchase grain on the world mar-ket. Moreover, with increasing living standards, more calories are being consumed as meat.[6] This increases the demand for grain to feed cattle. It takes 7–10 calories of grain to

---

[6] Other than by vegetarians, who e.g., comprise 30% of the Indian population.

produce 1 calorie of beef. These supply and demand factors reduce the overall economic impact of ethanol which, at present seems to be responsible for only a few per cent of the rise in world food prices. (Exact figures supporting this last estimate are lacking.)

Anthropogenic climate change

Of all the contentious and alarmist environmental causes, anthropogenic climate change and its proposed cure, or solution, may be the most pernicious. If the climate is indeed changing, and irrespective of whether human activity is to blame, the threat (if it exists) is not to be ignored. The anthropogenic climate change hypothesis has three major tenets:

- The concentration of CO_2 in the atmosphere has risen in the last 150 years from 25 Pa before the Industrial Revolution to 38 today, and is rising at about 0.1 Pa per year. This is a result of human activity.
- The climate of the world is changing, with many catastrophic consequences now in evidence and with much worse predicted for the next decades. The cause of these changes is a historically unprecedented rise in average world temperatures.
- The cause of the rise in world temperatures is the anthropogenic increase in atmospheric CO_2.

These claims have been investigated and given full support by the Intergovernmental Panel on Climate Change (IPCC), which was set up by the UN. The initial reports were backed by a gathering in Paris, in February 2007, of some 3000 scientists; mainly climatologists.[7] Add to this innumerable media reports and statements from environmental lobbies and celebrities riding the alarmist wave. Most prominent of the latter is the former Vice-President of the USA, Albert A. Gore Jr, whose commendable promotion of environmental causes earned him a Nobel peace prize.

The IPCC states (in its February 2007 report and 2008 updates) that it is 'unequivocal' that the rise in global temperatures has been caused by the greenhouse effect of elevated atmospheric CO_2. The IPCC predict a further 3–5°C rise within the next 50 years, if CO_2 should continue to rise unabated. It is also predicted that this temperature rise will have catastrophic world-wide consequences, including: melting of the polar icecaps, resulting in a rise in sea level, which will flood many low-lying areas, such as in the Maldives, Bangladesh, and the Netherlands;[8] drought in many currently grain-producing areas and flooding elsewhere; an increase in the number and force of storms and hurricanes;

---

[7] For the full report and latest IPCC updates see http://www.ipcc.ch/.

[8] Bangkok, the capital of Thailand, provides an example of misguided environmentalism. For a long time the media carried stories of the flooding of the low-lying city as a result of rising sea levels, caused by climate change. However, the sea is rising at a rate of at most 1–2 mm a year, whereas the town is subsiding at a rate of about 95 mm a year. Clearly, in this instance, local geology, not climate change, is the clear culprit.

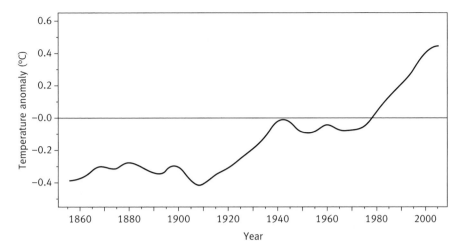

Fig. 9.6 Global air temperatures since since the beginning of the Industrial Revolution (~1850). The curve shows 5-year smoothed averages derived from IPCC data (see the IPCC URL in the References and Resources for latest updates). The zero temperature anomaly line is arbitrary.

increase in disease; expedited extinction of species; forced migration of populations seeking to escape famine and other ravages of the climate change; and many other woes.

Given such 'heavy artillery' a scientist has to by very brave to go against this consensus; but there are more and more intrepid souls, doing just that.[9] What exactly is being challenged? No one questions that the CO_2 concentration of the atmosphere has risen dramatically in the last century and a half (Fig. 9.4) or that the source of this CO_2 is mainly from the burning of forests and fossil fuels. It is also accepted that there seems to have been an average rise in world temperatures, albeit of the order of less than 1°C, over the last century (Fig. 9.6). Here the agreement ends. First, this 1°C rise is considered to be unexceptional. There have been many much larger swings in world temperatures over ancient and recent historic times. Moreover, there are many factors other than greenhouse gases which affect world temperatures. These counter-arguments are briefly summarized below.

One of the first to question the data of alarmist environmentalists was Lomborg (2001), whose work was noted above. He has recently written more specifically on the question of climate change (Lomborg 2008). Leroux (2005), himself a leading climatologist, questions the very foundation of the IPCC conclusions. He points out that the IPCC

[9] The Heartland Institute of Chicago prepared a list of 500 US scientists 'whose research contradicts man made global warning scares'. See http://www.heartland.org/policybot/results. html?artId=21978. For an updated list of '~30,000' names use the same URL, but with '=22387' at the end instead of '=21978'. But note that some of those listed have gone on record on the internet as having been misquoted.

bases its statements on simulation models of climate which are inherently unreliable. He questions its ability to make useful predictions for 50 years into the future, such as sensitivity studies of the effect of CO_2 on world temperatures. Two other recent books, all highly critical of the IPCC global climate change hypothesis, are by Singer and Avery (2007) and Svensmark and Calder (2007). They were written before the latest IPCC statements on the 'unequivocal' anthropogenic cause of the rise in world temperatures, via CO_2 emissions. Even so, it is very doubtful that their opinions would have changed. Some of the main arguments against the findings of the IPCC are as follows:

- *In the historic past there have been innumerable changes in climate, unrelated to human activity.* Ten million years after the global catastrophe at the end of the Cretaceous period (65 My BP) the Arctic region went through a warm period, with average temperatures above 20°C. This was correlated with a natural rise in greenhouse gases (both CO_2 and methane) and the placement of the tectonic plates at that time. Later there were numerous other periods of extreme cold (ice ages) with warmer interglacial stages (Chapter 7). We are still in the process of emerging from the last ice age, which ended about 11,000 y BP. If today we see melting, receding glaciers, this is still part of this process, although the rate of melt may have been accelerating in recent years. Another example of natural climate change can be seen in the Sahara region. Six thousand years ago it bore a tropical forest. It has been a complete desert only since about 2700 years ago. In the time of the Greek and later Roman empires North Africa was the breadbasket of the then known Western world. In historical times there have always been hotter, cooler, drier, and wetter periods. The Babylonian and later Biblical stories tell of long droughts and catastrophic floods. The stories are myths, but were based on long-term experience, thousands of years ago. The same climate variability continued in later times, with more accurate recording. A fascinating study of the last millennium, by Zhang *et al.* (2007), shows a correlation between recurring cool periods and warfare in eastern China. They found that during the cold periods crop production fell and warfare peaked. Between 800 and 1200 CE Europe went through what became known as the Medieval Warm Period. At this time both Iceland and southern Greenland were settled by the Vikings (or Norse) with agricultural communities. Prior to the 10th century the Mayas of Central America flourished and reached an estimated population of 13 M, an unprecedented number for the Americas in those times. Their decline and virtual disappearance as an organized civilization is thought to have been the result of recurring droughts in the 10th-century CE. The timing of these droughts correlates with the periodic cycling of solar radiation intensity (Hodell *et al.* 2001). From 1645 to 1715, and again for much of the 18th and early 19th centuries, there was a period which has been called the Little Ice Age.[10] Farming in Iceland became very limited. In Greenland it failed entirely and the

---

[10] This was related to the Maunder minimum of sunspot activity in the 17th century and then to the Dalton minimum in the early 19th century. Contrary to what one may assume, solar radiation is actually a little more intense during times of high sunspot activity.

colony declined and was abandoned. At that time, and for decades, the River Thames in London froze every winter (as vividly described in the novels of Charles Dickens). The Thames has not frozen in the last 140 years. From 1940 to 1970 the world went through a period in which temperatures not only did not climb but showed a cooling tendency (Fig. 9.6). In the 1970s climatologists issued warnings that we were entering a new ice age. Later, after a few warmer seasons, the 'global warming, climate change' Cassandras, took over. They stated that 1998 was the hottest year in US records, as was the last decade. A later data review showed that 1934 was hotter and five of the ten hottest years on record occurred before 1940. IPCC predictions of sea level rise *over the next 100 years*, have also been revised downward from the original ~90 cm to 43 cm.

- *The CO_2–world temperature relation is at best a correlation and not a very convincing one.* For example, as shown in Fig. 9.4(b) the rise in atmospheric CO_2 over the last century follows a smooth, monotonic curve. However, the rise in world temperatures (Fig. 9.6) shows a long flat and perhaps declining period between 1940 and 1975. There have been many publications (most famously the presentations of Al Gore) showing curves of historical atmospheric CO_2 levels and world temperatures over the several thousand years. They allegedly demonstrate the rise in temperature as a result of a rise in CO_2. There are, however, three problems with the conclusions drawn. The first is that the data from prehistoric times are derived from proxies (see above) and are therefore not very reliable. The second is that these are at best correlations: is the CO_2 driving the temperature, the reverse, or neither? Careful inspection of the twin curves shows that the rise in temperature usually precedes the rise in CO_2.

- *There are factors, other than the greenhouse effect which can modify world temperatures and climate.* They are 'natural', meaning quite unrelated to human activity. Moreover, the same doubting scientists calculate that the contribution of anthropogenic CO_2 to greenhouse warming can explain no more than about 15% of even the modest $<1°C$ temperature rise of the last century (Shaviv 2005).

As discussed above, in relation to the Little Ice Age, small changes in solar radiation intensity (of the order of 0.1%) have often been suspected of producing climate change on Earth. These direct effects seem to be small and the relation is uncertain (Rind 2002). Shindell *et al.* (1999) found that changes in stratospheric ozone correlated to the 11 year solar cycle and could amplify the effect of small changes in solar radiation intensity. A full analysis is given in Shaviv (2005). Shaviv (2003a,b, 2005), Shaviv and Veizer (2003) and Veizer (2005) have developed a hypothesis on how movement of the Sun through the arms of the Milky Way could modify the amount of cosmic ray radiation reaching Earth. Galactic cosmic radiation, as discussed in Chapter 4, produces nucleation and condensation of clouds, thus affecting the weather on Earth. These periodic movements correlate well with ice ages over the Phanerozoic aeon, a finding supported by Gies and Helsel (2005) and for the Holocene by Dergachev *et al.* (2006). Moreover, as discussed in Chapter 4, the solar wind (itself a form of low-energy cosmic radiation), whose intensity changes with the 11 year solar cycle, tends to attenuate the galactic cosmic radiation, thus modulating the cloud nucleation effect and affecting global weather patterns.

In brief, there is a healthy scepticism as to the very existence of global warming and climate change and, if they do exist, whether they are caused by the greenhouse effect of anthropogenic CO_2 or by extraterrestrial factors. There are many reasons why we should not burn forests and fossil fuels but, even if we manage to reduce atmospheric CO_2 this may have little effect on world temperatures. It may be asked: 'What's wrong with reducing atmospheric levels of CO_2 as a wise precaution?'. The reason not to is rather simple. Lowering CO_2 may not affect the rising world temperature and doing so uses resources that could be better utilised. A number of other possibilities for cooling the world have been suggested. Most have proposed some means for increasing the global albedo, such as releasing particles and aerosols into the stratosphere. So far none have appeared with a thorough global and environmental impact assessment study.

The above subsections reviewed just four cases of environmental extremism. Good, well-meaning intentions, but poor data. By the time you are reading this they may well be passé, having been replaced by the latest 'cause of the week'. But healthy scepticism when facing the latest environmental alarm should not make us insensitive to the effects of humanity on the biosphere. They are very great and often extremely damaging. It is of great concern that catastrophic climate change may be taking place, and even accelerating, as a result of completely natural causes, unrelated to humans. Because we may be blameless does not mean that we should passively accept the situation, without seeking solutions.

As discussed above, many of the 50 and 100 year projections of the climate change thesis are being challenged. Even so, and despite the fact that the present rise in world temperatures is relatively moderate, there are some signs of climate change which deserve very close monitoring. For example, in the summer of 2008 the fabled North-West Passage (for boats sailing north of Canada, from Europe to Asia) became almost clear of ice. At the same time there has been an acceleration in the melting of Greenland glaciers. In the South, NASA satellites have recorded the break-up of large patches of the Antarctic ice shelf. Add to this the meteorological finding that since the beginning of the 20th century there seems to have been an increase in the frequency and intensity of extreme meteorological events (storms, hurricanes, drought, and flooding). Together they *may* be signs of climate change in the direction of global warming.

One fearsome outcome of polar warming is that it may cause a feedforward runaway heating effect. This could be the result of a reduction in the area of snow and ice which reflects solar radiation and an associated increase in the area of open sea water which absorbs radiation. An even more frightening scenario is that if the Arctic permafrost were to melt it would release vast volumes of methane gas. This, as noted above, is a highly potent greenhouse gas which could further accelerate runaway global warming.

Assuming that none of the above doomsday scenarios comes about, we are still left with the problem of caring for the ever-increasing human population.

How many people can the world support?

As shown in Fig. 9.3 the population of the world is expected to reach about 9–10 G within the next 50 years. As discussed above, today (2009) there are world food shortages even outside those areas of chronic food shortage such as in parts of Africa. These shortages may soon be met, as the economics of supply and demand put more resources into food production. But are such corrections sustainable over the long term? Will we always be able to produce enough food for the growing population? Moreover, we would like everyone to attain the food standards now enjoyed by only 20% of the world's population. Is such a target attainable?

One of the first to ask these questions was the economist Thomas Malthus, who published his famous essay on population in 1789, with revised editions appearing over the next quarter of a century. The basis of his thesis is that populations tend to increase geometrically, while food production increases arithmetically. Consequently and inevitably, the amount of food available per person will decrease. The result is increasing malnutrition, infant death, disease, and wars, which tend to 'correct' the imbalance. He put the blame for this situation on the 'working classes' who, in their irresponsibility, reproduced without any concern for their ability to support their families. According to Malthus the food supply, or lack of it, controls the population size. Malthus, in his later writings recognized the human talent for generating new technologies, thus breaking the 'arithmetic food increase' impasse; but he thought that this too must ultimately come to an end. He was much criticized for his social views and by later economists who thought that population growth drives agricultural innovation and advances productivity. Malthusian ideas are still being discussed, whatever the underlying mechanisms and social innuendos.

In the time of Malthus (the early 19th century) there was a world population of about 1 billion, most of whom were chronically hungry. Today the population of the world has passed 6 billion, while global-scale famine has been practically eliminated. Moreover, the standard of living of a fifth of the world's population is unprecedented. They eat more food, of a higher quality, than ever before.

Whether today's relatively benign food situation can be maintained when the population of the world reaches 9 G in about 50 years from now (Fig. 9.3) is of paramount concern. The basic considerations were set out by Cohen (1997). He believes that the answer lies between the 'cornucopians and the doomsayers' and that we should 'live in neither panic nor complacency'. Solutions are possible if we act wisely. Fedoroff and Cohen (1999) introduce and review an important symposium which dealt with many aspects of this problem, and which reached similar conclusions.

Three books dealing with these all-important questions have appeared recently, each with a different approach (Conway 1997; Evans 1998; Smil 2000). They were reviewed, together, by Siedow (2001). Siedow first set out a strong case for the introduction of genetically modified organisms (GMOs) into the agriculture of developing nations. To summarize the argument briefly: shoppers in developed nations are so spoiled by the availability of low-cost, high-quality food, that they can afford not to accept any food for which they have the slightest reservation. Not so in the developing countries, where the introduction of GMOs could make a major contribution, perhaps a second Green Revolution. Siedow reviews the rather different approaches of the three authors. At least one of them (Conway, an economist), also sees GMOs as a powerful potential factor in the struggle to avoid world-wide food shortages. The bottom line is very much the same as that arrived at by Cohen (1997) and by Fedoroff and Cohen (1999). The problem of feeding the future world population can be solved if agricultural technology is allowed to advance and at the same time investments are made in the health and education of the people using it. Add to this the marketing and social revolutions required in many regions. For example, it has been estimated that in India farmers receive only one-fifth of the price paid by the consumer.

A note on land and energy resources

The section above addressed the problem of food, and the problems of fresh water supply have been reviewed in relation to pollution. However, the availability of agricultural land has not been discussed directly. Although of great importance, with many local problems of degradation and loss of land to urbanization, on a global scale there is still much land available. Given sufficient fresh water for irrigation (easy to say, but still a major problem) there are vast arid and semi-arid marginal lands which could be put into productive agricultural use.

The exhaustion of fossil fuels is a world problem. Today it is mainly a question of huge demand which goes unsatisfied by the cartel of oil producers, interested, of course, in high prices. Tomorrow there will probably be a reduction in supply as sources dwindle. In the absence of new discoveries of major oil fields, or the development of clean technologies to convert coal and oil shale to liquid fuel, the situation will become super-critical in 20–50 years from now. Not only will the world population be heading for a 50% increase, but the percentage of people who use vehicles and oil-driven machines, whether for private, industrial, or commercial activities, will increase. This is already happening in fast-developing, highly populated India and China. It has been calculated that if the percentage of the population using cars in India today was the same as in the USA, the world demand for gasoline would be doubled. This is just vehicle energy use. The demand for electricity, for industrial, commercial, and domestic use, is expected to increase proportionally.

The rapidly increasing demands, diminishing supplies, and rising cost of conventional energy will no doubt drive the development of alternative sources, which today are too expensive. These include solar, wind, geothermal, and nuclear energy. Since the Chernobyl accident discussed above, nuclear power stations (utilizing fission technology) have become very unpopular and very few new power stations have been built. Even so, there are still many which operate safely. France, for example, derives 78% of its electricity from nuclear power stations. (Let's hope that these last two statements do not belong in the realm of 'famous last words'!)

Eventually the problem of the control of nuclear fusion (of hydrogen) in commercial power stations will be solved. When it is, the world will enter a new era of cheap and unlimited energy. The impact will be in all areas. For example, ecological clean-up processes, such as stack emission scrubbing, will become economically worthwhile. (However, stacks may no longer be necessary, as factories turn to low-cost electricity.) Low-cost desalination of sea water may also be realized. Apart from satisfying industrial and domestic needs, such water may even become inexpensive enough for agriculture, at least in greenhouse-type closed systems if not in open fields. Such systems, using technologies already in existence today, could be set up on any marginal lands. They would contribute to satisfying the future world food demand. Moreover, if used sparingly in water-conserving irrigation systems, low-cost desalinated water may even be used in open-field agriculture for high cash crops.

It is satisfying to conclude this chapter on the above optimistic note. Physicist acquaintances tell me it is still a far-off dream.

Summary

- Humans, *Homo sapiens sapiens*, first appeared in Africa about 200,000 y BP. They evolved from the genus *Homo* which appeared some 5 My BP. Humans are distinguished by an upright posture, large complex brains, and tool-grasping hands.
- Humans dispersed from Africa to the rest of the world some time between 60,000 and 100,000 y BP, reaching the Americas via the Bering Land Bridge, about 20,000 years ago.
- The human population was sparse until the end of the last ice age (~11,000 y BP). It reached millions in the early Middle Ages. After the 17th century, numbers began to rise exponentially to today's 6 G, with a prediction of ~9 G in around 2050.
- Humans damage the biosphere in many ways:
  - Air pollution: includes CO_2 from burning forests and fossil fuels; methane from cattle, sheep and rice padies; ozone (toxic to life in the troposphere); CFCs, which destroy UV-absorbing ozone in the stratosphere; aerosols which increase Earth's albedo; sulphur gases which produce acid rain; and many other toxic gases and dusts.
  - Water pollution: all the above air pollutants, and more, reach the oceans and fresh waters. The oceans also receive effluents from coastal and inland communities. Fish populations have been drastically reduced by over-fishing. Global shortage of fresh water frequently makes its pollution critical.
  - Radiation: pollution by radiation is relatively minor today but, tomorrow, should there be a nuclear war, humanity and much of the biosphere could be destroyed.
  - Extinctions: few complex, multicellular species survive for more than a million years. In recent years there seems to have been a rapid acceleration in human-related species extinction. Most biologists believe the numbers are approaching catastrophe; others claim that present extinctions do not exceed the normal extinction rate.
- Four examples are given of exaggerated and often incorrect environmental 'causes':
  - The tropical forests are stated to be the lungs of the world, supplying oxygen and reducing CO_2. Important as they are, until burnt they have no effect on the CO_2 and oxygen content of the atmosphere.
  - High CO_2 in the atmosphere is nearly always mentioned as a negative factor. In fact it increases the growth of most wild and agricultural plants and makes them more resistant to drought and salinity.
  - Ethanol is promoted as an energy source for cars, as it is thought to save fossil fuel and reduce CO_2 emissions. When all factors involved are taken

into account it is found that neither claim is correct. Moreover, it diverts food resources from the poor to provide car fuel for the wealthy. Other statements, to the effect that the recent introduction of ethanol technology is the reason for the rise in world food prices, are also exaggerated.

- Anthropogenic atmospheric CO_2 is claimed to have increased world temperatures, resulting in catastrophic climate change. This thesis is backed by an extensive UN report. Many others question the basic findings, arguing that (1) the climate models used do not have the ability to make the predictions claimed, (2) that there have been numerous historical instances of greater temperature changes than found in the last century, (3) that the contribution of elevated CO_2 to temperature rise is minor, and (4) that there are other, extra-terrestrial factors which modify global weather, such as small changes in solar and cosmic radiation, which are modulated by the varying position of the Solar System in the galaxy.

- Malthus predicted inevitable starvation, resulting from exponential population growth and arithmetically increasing food production. This has not proven correct, despite the population having increased from 1 G at the time of Malthus to 6 G today. Whether this currently benign situation will continue when the world population increases to 9 G is uncertain.

- Potential agricultural land is not in short supply on a world-wide basis. There is much semi-arid land which could be put into use, if (a big if) supplied with irrigation water.

- Today the demand for energy, especially liquid fossil fuels, is rising rapidly and exceeds supply. Sources of liquid fossil fuel may be exhausted within 30 years. Alternative energy sources are becoming attractive as fossil fuel prices increase. These include solar, wind, geothermal, and nuclear fission energy. The latter does not produce CO_2 or other 'usual' pollutants but has severe problems of safety, security, and radioactive waste disposal. Hydrogen fusion power is potentially the most promising alternative energy. It produces no pollutants and its raw material (water) is essentially unlimited. However, 50 years of research have not solved the engineering problems of its containment and control. Problems of safety and security would remain, as for nuclear fission power stations.

10
In search of extraterrestrial life

The search for extraterrestrial life, and especially intelligent life, is one of the most exciting scientific enterprises of our time, although in fact it is a very ancient quest. It is the central theme of astrobiology, of which the astrobiology of Earth is just one aspect. Only in the last few decades has technology attained a level which has enabled the commencement of a science-based search for the existence of conditions amenable to life outside Earth. This includes a search for signs of other intelligent civilizations, which may be using EM radiation or other means of communication.

However, in relation to the size of the cosmos and the number of possibly promising sites for life, the present search is still in its infancy. Note that we are still looking for 'conditions amenable' to life. Clear evidence of life or an intelligent signal from an extraterrestrial source has not yet been obtained. We still do not know whether or not we are alone in the universe.

Some of the subjects treated here were first discussed in Chapters 2 and 3, where the universe was reviewed from the point of view of biology. The initial search for extraterrestrial life is limited to 'life as we know it'. This is a qualification of convenience. There is no way to be sure that life elsewhere would not be very different from that on Earth, having developed under vastly different environmental conditions. Even if we find a planet with an environment similar to that of our biosphere, any life which developed there may be quite different from that with which we are familiar. Only one basic life form developed on Earth. We have no reason to believe that it was the only one possible. Moreover, we are still having great difficulty in trying to understand and define 'life'; so we don't know exactly what it is we are looking for (Chapter 1). If a radio signal is sent our way will we be able to understand it? The senders may be using entirely different language syntax and mathematical formulations.

If life, certainly intelligent life, was discovered elsewhere, and if two-way communication could be established (today, almost impossible: read on) this would be one of the greatest cultural events in human history. It is consequently not surprising to find that such possibilities have been discussed by almost every human civilization, primitive or advanced. This has been in the form of religious and philosophical discourse, rigorous science, science fiction, and everything in between.

The relentlessly increasing human population is rapidly crowding the biosphere and depleting the raw materials on which human life on Earth depends. This includes arable soil, water, and energy. Moreover, much of what is available now is being polluted.

It is therefore hardly surprising that science fiction frequently contains the theme of humans escaping from Earth and settling on planets of other star systems. It is proposed that this would parallel the colonization by European nations of the Americas and other less developed regions of Earth in the 16th to the 20th centuries. How realistic are such speculations?

A short history of the origins of astrobiology

When I consider thy heavens, the work of thy fingers, the moon and stars which thou hast ordained: What is man that thou are mindful of him?
Psalms viii 4, 5 [attributed to the biblical King David, ~1000 BCE]

Twentieth-century scientists were certainly not the first to wonder at the heavens, neither was King David. More than 4000 y BP the ancient Babylonians observed and recorded the place and movements of the heavenly bodies as seen on dark desert nights. This included stars, planets, the Moon, comets, and eclipses. At about the same time the ancient Chinese began some of the first reliable astronomical observations and recordings, such as of the solar eclipse of 2136 BCE. Less happily (for science) the Babylonians were among the first to dabble in astrology. They purported to predict great events, such as wars, the death of monarchs, and famines, from the positions of the stars and planets. Early Hindu Vedic writings in India also showed a similar interest. At that time there were cultural links between ancient Mesopotamia and India, which may have encouraged this convergence.

Some 1500 years later the ancient Greeks developed not only astronomical observations and arithmetical predictions, but also conceptual hypotheses. In retrospect, Democritus, who lived from 460 to 370 BCE in Abdera, Thrace (the north-eastern part of modern Greece), was one of the most amazing (Fig. 10.1). He travelled in the ancient world and learnt from Persian astronomers. His teachings are of great interest to astrobiologists, although he is best known as one of the first to propound the 'atomist' theory. Steven. J. Dick (1982), a historian of the concept of extraterrestrial worlds, argues that there was a direct logical link between the atomist theory and belief in other worlds.

The books of Democritus and his immediate colleagues have not survived. We know of their teachings from later Greek and Roman authors. It is from them that we learn of his extraordinary perspicacity. He realized that the Sun was just a star and that what appears to be the 'fog' of the Milky Way is composed of innumerable distant stars. He understood that the planets revolved around the Sun and that Earth is a planet. He further theorized that there would be planets around other stars, 'some of which would have one or more suns and moons'. Modern astronomical observations indicate that the latter hypothesis is possible. According to Democritus some planets would be arid and lifeless while others would bear life similar to, but not necessarily identical to, that on Earth. His teachings agree well with 21st century astrobiological thought.

Aristotle (384–322 BCE) soon followed Democritus and the early Greek astronomers of the atomist school. Like Democritus he too was born in Thrace, on the Macedonian

Fig. 10.1 Democritus (460–370 BCE) the first(?) astrobiologist. Picture of a Roman era sculpture.

coast. Aristotle studied under Plato in Athens. The school of philosophy and science which he founded formulated the entire framework of classical science. Until the Enlightenment, which followed the Renaissance (see below), their writings were considered almost sacrosanct. Sadly, many if not most of the teachings of Aristotle's school were later shown to be incorrect.

The Aristotelian school firmly rejected the idea of life existing anywhere but on Earth. Even so, some later Roman thinkers disagreed. For example, a belief in other inhabited worlds is expressed in the writings of the first century BCE poet and philosopher Lucretius, a follower of the atomist school, from whose writings we know many of the concepts of Democritus.

The apex of Aristotelian astronomy was reached in the 2nd century CE. This was in the school of Ptolemy, in the Greek colony of Alexandria, on the coast of present-day Egypt. The school of Ptolemy put the Earth at the centre of the universe, and produced sophisticated and complex mathematical models to account for the apparently errant movements of the planets. Their model held sway until the 15th-century CE, when the work of the Polish astronomer Copernicus, followed by the 16th-century Italian Galileo, put the Sun at the centre of the Solar System. Galileo was one of the first astronomers whose observations were made with telescopes. They were of his own design and construction.

One unfortunate product of Greek astronomy was astrology in the guise of personal horoscopes. Predictions of events were based on the positions of the planets and stars on one's birth date. This superstition would hardly be worth mentioning were it not for the unfortunate fact that a belief in astrology and horoscopes has persisted through the ages—just see today's popular press and television or listen to radio broadcasts!

There were always those who doubted the veracity of astrology. For example, Maimonides, a great polymath of the 12th century, was a student of Aristotelian science. Even so, he regarded astrology with disbelief and disgust. He was also an internationally famous rabbi. As such he taught that astrologers and their followers were no better than idol worshippers. However, others, for example even the great astronomer Kepler who formulated the mathematics of planetary motion in the late 16th century, had an interest in this occult pursuit. William Shakespeare, another genius of this same period, thought otherwise:

Astrology: This is the excellent foppery of the world: that when we are sick in fortune-often the surfeits of our own behaviour-we make guilty of our disasters the sun, the moon, and stars, as if we were villains on necessity, fools by heavenly compulsion, knaves, thieves, and treachers by spherical predominance, drunkards, liars, and adulterers by an enforced obedience of planetary influence...

King Lear, I, ii

It would perhaps be best to dismiss astrology with a single sentence or not to mention it at all, but this would ignore a serious problem in education. Faced with the wild predictions of tabloid astrology most people, even those with only a modest education, tend to jump to the opposite extreme of belief. In correctly disparaging astrology, they conclude that the heavenly bodies have absolutely no effect on our lives. Contrariwise, the study of the astrobiology of Earth, as unfolded in these short chapters, shows how the various bodies and forms of energy in the cosmos have in the past affected and indeed effected the environment of our biosphere. Moreover, they continue to do so today. Even so, the dates of our birth and the position of the stars, planets, and comets do not determine the outcome of wars, whether we shall be successful in our investments, or if we shall meet a new love next Tuesday! Unfortunately, astrology and horoscopes are without merit. At best they provide some light entertainment. At worst they promote false hopes or fears in the gullible. They persist simply because we would so much like to predict the future. Alas, this is not possible.

While criticizing some of the conclusions of ancient astronomers, especially in their astrological manifestations, we must always bear in mind that their work was based on observations made with the unaided human eye—no Hubble telescope circling the Earth above the optically disturbing atmosphere, nor even the simple Earth-bound telescopes of Galileo. They made careful observations, kept records, and drew sophisticated, mathematically precise (if sometimes totally inaccurate) conclusions. Many of their calculations, made without computers or slide rules, were truly astonishing. For example the Babylonian astronomer Kidinnu (4th-century BCE) calculated the diameter of Earth to within just a few per cent of the present figure (12,756 km) and estimated the length of the year to within only 4.5 minutes of its true value!

The Renaissance, a rebirth of Western science and culture, which commenced around the 15th century, began with Aristotelian science. This 'classical learning' had been kept alive and elaborated during the Dark Ages by the flowering of Arabic culture after the 7th century. It included the preservation of Ptolemy's astronomy and the further development of astronomy, chemistry, mathematics, and medicine. Belief in or denial of the possibility of the existence of worlds other than Earth became mainly a scholastic, religious argument. Tycho Brahe and Giordano Bruno, in the late 16th century, and Bernard le Bovier de Fontenelle in the 17th were the main advocates of the plurality of worlds. For this and his other 'heretical beliefs' Bruno was tried by the Venetian Inquisition and burnt at the stake in Rome in 1600.

In the 17th and 18th centuries Western culture progressed through a period known as the Enlightenment in which rationality was preferred to tradition and untested beliefs. Although it included the beginning of experimental science, there was still a measure of speculation. This included the belief of many savants, such as Henry More in England, de Fontenelle in France, and Christiaan Huygens in Holland, that stars were like our Sun and could be the centre of planetary systems which support life. It is interesting to note that in the early 18th century Isaac Newton, who speculated on the existence of other worlds, had no problem with reconciling these ideas with his deeply religious beliefs. There were many who debated the plurality of worlds thesis in the 18th to the 20th centuries, but there was little added (Dick 2001). Religious and metaphysical arguments make them difficult and generally irrelevant to the modern scientist.

The possible existence of intelligent life on other planets of our Solar System and on the Moon has long been a subject for speculation. This goes back to at least the Greeks of the 5th and 6th centuries BCE. The arguments continued right through the Enlightenment. Proponents included Kepler in the early 17th century. Perhaps the last of the astronomers who believed in the past (if not the present) existence of intelligent extraterrestrial life in the Solar System was the American Percival Lowell (1855–1911). Lowell thought he had evidence showing that in times past Mars was a moist planet which supported life and an advanced civilization. According to Lowell, as Mars gradually began to lose its atmosphere, water became scarce and the Martians built a system of canals to transport water from the poles to the regions of their settlements—a not illogical scenario. Lowell published numerous maps of Mars showing these putative canals, based on his telescopic observations (e.g. Fig. 10.2). Today, with our modern Earth-bound and space telescopes and the pictures returned by Mars-orbiting probes, we know that no such canals exist, extant or vestigial. What Lowell drew was what his eye and brain interpreted from the surface markings, as seen with the telescopes of his

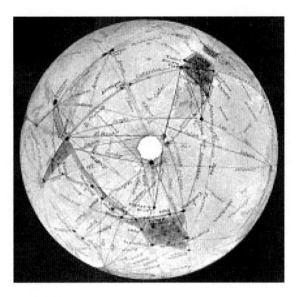

Fig. 10. 2 Martian 'canals' as drawn by Percival Lowell (1855–1911).

day, and from his preconceptions. Today, no one expects to find extraterrestrial intelligent life in the Solar System, extant or extinct.

The search for extraterrestrial life today

There are three ways in which we could search the universe for life and for the environmental conditions which have the potential to support life. The first, and certainly the most attractive, is by space travel. The second is to use automated probes. The third mode of exploration, and currently the only practical method of investigating the extra-Solar System universe, is by spectral analysis of the radiation reaching Earth from such systems. These include the familiar visible spectrum, other wavebands of the EM spectrum, and, occasionally, particles of matter. The latter may range in size from protons to meteorites and comets.

In order to search for life, as discussed in Chapter 1, we must first define it. There is great difficulty in formulating such a definition, even for life as we know it on Earth. In considering search parameters it was noted that, although our galaxy and certainly the entire universe are vast, regions where Earth conditions may exist are quite limited. Eighty per cent of the volume of the galaxy must be ruled out a priori (Chapter 2). Life which developed under environmental conditions different from those with which we are familiar on Earth may be so esoteric as to defy detection. This is one of the major problems of general astrobiology.

Consequently, and it cannot be repeated too often, the initial search for extraterrestrial life is informed, coloured, and to a large extent limited by our knowledge of life on

Earth. One of the most detailed summaries of the criteria for this search can be found in J. I. Lunine's (2005) astrobiology textbook. Some of the main features of the search criteria are given in Table 10.1. Today, this search is mainly confined to the Solar System. As discussed above, there are a number of Solar System sites which may have niche environments which could support at least simple life forms, particularly of the type categorized under Earth conditions as extremophiles (Chapter 8).

With respect to the search for life outside the Solar System, we are currently only at the very first stage. In the last decade this search has received much encouragement from the detection of numerous extra-Solar System planets. In the coming decades we can expect to have space-based telescopes capable of analysing their atmospheres, where present. Specific wavebands of the EM spectrum, such as the near-infrared (NIR) may indicate the presence of organic molecules of possible biological origin. The ESA is currently planning a NIR interferometer-carrying satellite, which has been given the name Darwin.

Why not fly there and see for ourselves?

Humans have one huge advantage over automated probes: our biological computer, otherwise known as a brain. It has pattern recognition abilities superior to those of the best present-day electronic computers. Moreover, in spaceflight the unexpected is the norm. Humans are often capable of improvising ingenious solutions to these unanticipated space and planetary surface problems. However, long-distance piloted spaceflight is highly problematic. Some of the main limiting factors are summarized in

Table 10.1 Criteria used in the search for extraterrestrial life

- Presence of water—and for a start, temperature and atmospheric pressure conditions—that could support water in the liquid phase.
- Cells—meaning membrane-enclosed structures, with recurring inclusions which may support information storage and metabolism, in which the internal milieu is maintained with a different composition from that outside the cell.
- Structural complexity—showing dissimilar structures at increasing levels of development. This seems to be a feature of higher, multicellular life forms.
- Stereoisomeric structure—composed of molecules with either a constant left- or right-handed chirality.
- Organic (carbon-based) chemistry which may exhibit considerable disequilibrium.
- Presence of RNA, DNA, and certain stable carbon, nitrogen and sulphur isotopes.
- Presence of amino acids—the building blocks of proteins. Their presence may indicate life which existed in past aeons. RNA and DNA molecules are not so long-lasting.
- Atmospheric composition—the presence of certain reactive gases, e.g. oxygen and methane may be indicative of ongoing life processes. However, other non-biotic sources of these gases are possible[a].

[a] For example Titan, a satellite of Saturn, has large quantities of liquid methane on its surface which are almost certainly not of biogenic origin.

Table 10.2. In addition to the factors listed in that table there are very real engineering problems, such as the currently inadequate propulsion systems required for extra-Solar System flights, whether or not they carry a human crew.

It is the combination of great distance, relatively slow speed of travel, human frailty, and short life span, which together puts travel to distant planets of the Solar System and certainly to other star systems (not to mention other galaxies) well beyond present human capability. To this must be added cost. Within our Solar System sending humans to Mars may be just within human physiological and present technological capabilities. However, even the wealthiest of nations may not be prepared to allocate the enormous resources which would be required.

We think of the speed of light as being extremely fast, but in the context of the distances in space and the short life span of humans it is not. It is also highly unlikely that, with present-day engineering, a spaceship could attain more than about 10% of this speed, even after years of continuous acceleration. It is estimated that, with present spaceflight technology, a trip to the nearby planet of Mars and back would take about 2.5 years. This would include just a few weeks on the Martian surface.

Should greater flight speeds, close to the speed of light, be attainable there may be an *apparent* solution to the astronauts' short life span. This is a result of time dilation and distance contraction, a prediction of the theory of relativity. The astronauts would age more slowly. However, should they ever return to Earth, they would find that the Earth mortals who had despatched their flight had died generations ago.

There has been limited progress in developing means for overcoming the health hazard of exposure to radiation in protracted spaceflights within our Solar System. It is possible that small safety areas with heavy shielding could be built into the spacecraft.

Table 10.2 Limitations to piloted travel in space

• The huge distances involved, for example:	from Earth to Moon ~386,000 km from Earth to Mars ~5.6×10^7 km (when in apposition[a]) from Earth to nearest star ~4.3 ly from Earth to edge of the Milky Way ~10^5 ly from Earth to the Andromeda galaxy ~2×10^6 ly
• Spaceship travel speed, ultimately limited by the low speed of light, which is: only 300,000 km s^{-1}	
• The short lifespan of the human species:	<120 years (and <30 years for the working life of an astronaut)
• Health problems in extended space flight, resulting from:	exposure to radiation (such as solar flares) long-term micro-gravity long-term physical inactivity boredom and psychological and social deprivation

[a]The closest it gets to Earth (which occurs when both planets are in opposition, i.e. on the same side of the Sun).

The latter would be used as shelters at times of heavy incident radiation, as experienced during solar flares. The onset of solar flares could be detected from Earth, and early warnings radioed to the astronauts.

Calcium loss from the skeleton and reduced muscle tone are among the main health problems resulting from prolonged exposure to micro-gravity. They may be partly overcome by exercise and dietary supplements. This has been studied in astronauts who have undergone spaceflights which lasted for about a year. There are other health hazards with no present solution, such as a reduction in lymphocyte blood count. To date, even after relatively short year-long spaceflights and despite the best counter measures currently available, astronauts have returned to Earth in poor health.

The ability to live for long periods of time in a closed metal and plastic box, with just a very few other people, requires exceptional personality traits. Such traits may be quite different from those considered desirable for life in a normal social environment on Earth. Boredom, sexual deprivation, and the development of undesirable relationships among the astronauts may result in severe problems (Simpson 2000). The results of a discreet NASA panel of inquiry in 2005 pointed to the psychological and social complications that could occur if the subject of human sexuality is ignored.

The first humans landed on the Moon, our nearest neighbour in space, in 1969. Not unexpectedly, for such a dry body with no atmosphere, no signs of life were discovered. Piloted travel to sites in the Solar System more distant than the Moon or Mars would be far more difficult. For example, Saturn's satellite Titan, which automated probes showed to have lakes of liquid organics, may have conditions similar to those existing on early Earth. Present conditions on Titan could perhaps throw light on the emergence of life on Earth. What biologist would not want to pay it a visit? But it is more than 20 times the distance of Mars from Earth.

Creating a micro-biosphere, to enable humans to travel in space

A major problem with piloted human spaceflight is how to carry sufficient supplies to support biological needs. These include food, water, and oxygen. Astronauts also produce liquid and solid waste, from washrooms, perspiration, urination and defecation, food and packing materials, etc. Humans respire, consuming oxygen and exhaling CO_2. All these requirements are now met with supplies taken up from Earth. These include food, oxygen, and expendable chemicals, such as CO_2 absorbers, for treating the effluents and pollutants. A return trip to Mars would require about 5 tonnes per astronaut.

A long-term flight with, for example, six astronauts on a 2.5 year expedition to Mars, would consequently require about 30 tonnes of provisions. A rough guide puts the cost of launching 1 kg (of anything) into space at about $10,000–50,000. The cost of life-support provisions for a trip to Mars would therefore be in the range of hundreds of millions to billions of \$US. Even within the overall cost of spaceflight this is a very large component, which invites a search for alternatives.

On Earth we live in our biosphere which recycles most animal and plant wastes. For example, animals, insects, plants, and microbial life respire. In so doing oxygen is taken up and CO_2 is released. This is more or less balanced by the activity of plants in the oceans and lakes and on land. In the course of photosynthesis plants absorb (fix) CO_2 and release oxygen as their (to them toxic) waste product. At the same time plants transpire pure water. Pure water evaporated from oceans, lakes, and rivers, together with the transpiration of land plants, is recycled in the biosphere. Similarly, human and other animal waste is recycled in the biosphere by microbial and geological processes. The biosphere's biological cycles generally balance out. (When this process is less than perfect, as has frequently been the case in the past, periods of high temperatures or ice ages may occur.)

The energy requirements of a large part of the non-photosynthesizing heterotrophic microbes, insects, and fauna are obtained directly or indirectly from plants. Simple pro-karyotic life supported by energy-rich molecules does not need photosynthesis. While the latter constitute a considerable fraction of the biomass of the Earth's biosphere, microbial life based on energy-rich molecules is of little quantitative importance on the surface of Earth or in the context of a spaceship.

It has long been suggested that the expense of carrying huge quantities of provisions on long-term spaceflights could be at least partly offset by setting up a closed biological loop (or micro-biosphere) within the spaceship. This would centre on an area devoted to growing plants. The concept of a controlled environment life support system (CELSS) or a bio-regenerative system is shown diagrammatically in Fig. 10.3.

The CELSS concept was pioneered more than four decades ago in the Soviet Union. They reached the level of constructing (on Earth) an airtight greenhouse system where a selection of food plants was grown. Cosmonauts were able to live for a few weeks at a time in these enclosures, eating the plants and having their respiration balanced by the plants' photosynthesis. Much further work on different aspects of such systems, such as optimizing plant growth conditions or dealing with waste, has been carried out in the USA, Europe, Japan, Israel, and elsewhere. However, bringing CELSS from an ideal concept to practical reality is fraught with huge problems.

The first challenge for CELSS is to minimize the area required for plant growth. Under ideal conditions of high light intensity and optimum water, nutrition, and CO_2, it has been calculated that only ~4 m^2 of floor space, in a growth chamber of 6 m^3, would suffice to produce all the oxygen, food calories, and CO_2 absorption for one crew member. This would mean about 20 m^2 and 30 m^3 of costly spaceship area and volume, respectively, for a five-member Mars expedition crew.

The second major CELSS problem is the energy required for photosynthesis. Maximum potential plant growth requires an incident light intensity of some 2000 μmol photons m^{-2} s^{-1} of photosynthetically active radiation (PAR; in the 400–700 nm waveband).

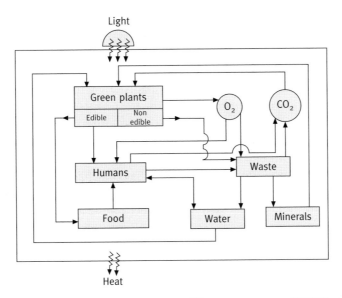

Fig. 10.3 The concept of a controlled environment life support system (CELSS) or bio-regenerative system, for extended spaceflight.

Even assuming the use of highly efficient light emitting diodes (LEDs) it is estimated that about 1 kW m^{-2} of electricity would be required. Photosynthesis has only a low theoretical efficiency for light use conversion ($\sim$12%). Even under the best growth conditions only about 4% of the incident radiation is converted into plant material, of which only part is edible. The lamp heat, from electricity not converted to radiation and from non-PAR photons and unused PAR (which also ends up as heat) must be dissipated. Cooling systems demand a further expenditure of energy.

Theoretical work in Japan has shown that if the spaceship is within the Solar System and relatively close to the Sun it may be possible to introduce sunlight to the ship's plant growth chamber via a system of collection mirrors and fibre optic channels. Farther from the Sun, electrical energy must be used to drive and cool the lamps. At the present time, in the absence of a safe and reliable nuclear reactor for use on board a spaceship, it is unclear where this will come from. Moreover, the lamps themselves have a limited lifetime and must be regularly replaced (a lesser problem if LEDs are used).

There are numerous other problems with CELSS which must be overcome before such systems could become practical for spaceflight. Among these are the selection and cultivation of plants which grow more or less normally under conditions of micro-gravity and which could produce a balanced (if vegetarian) diet. Micro-gravity is what would be encountered on a spaceship. Gravity would be less of a problem on CELSS systems set up on the Moon, which has one-sixth of the gravity of Earth, or on Mars with one-third of Earth's gravity.

Another major problem is the sterilization and breakdown of bio-waste from the crew's quarters and the inedible part of the plants. They must be reduced to a form which can be used as plant fertilizer. Such a waste treatment facility requires an in-depth knowledge of the microbial life in the spaceship. Evolution of microbes is much faster than that of higher life forms. On a spaceship they may transmute to undesirable forms. This demands monitoring and control—an extremely difficult on-board task.

First and foremost among the problems of CELSS is reliability. The growth of plants in controlled environment chambers on Earth is notoriously unstable. Systems will often work well for a few months and then collapse because of a mechanical or electronic control problem or the emergence of some insect, fungal, or bacterial disease. In space, any of these factors may defeat all the means used to test, isolate, and clean the systems before takeoff from Earth.

A positive side of growing plants in space is the psychological factor. Modestly sized plant growth systems may not be large enough to fulfil all the aims of a CELSS. They could, however, produce small quantities of fresh plants, such as lettuce. The mere occupation with day to day cultivation, involving contact with growing plants, is believed to be highly beneficial to people otherwise exposed for years to a sterile and artificial environment. Consider a fresh salad, served after months of toothpaste-tube food, or a small fragrant rose placed at the centre of the mess-room table!

The bottom line for acceptance of even a well-tested and reliable CELSS system depends on the expedition flight time. The take-off weight of the CELSS itself, including support equipment and spares (such as replacement lamps), must be less than that of the alternative, namely packed provisions and waste treatment materials. To date there have been only a few attempts to make such a pay-back calculation. One NASA report (going back some two decades) estimated a minimum time of 8 years to meet the breakeven point. It was considered by many to be highly optimistic. This means that CELSS would not be an attractive technology for human spaceflights of less than about 10 years' duration. There are currently no such expeditions on the books.

So, have all the research efforts into CELSS been wasted? It depends on who you ask. Research on maximizing yields under highly controlled conditions of light, temperature, CO_2, humidity, etc. may help to increase agricultural yields on Earth. To many researchers the importance of CELSS research on micro-biospheres is that it furthers our understanding of how the Earth's biosphere maintains its currently human friendly, if easily perturbed, equilibrium conditions.

Technology developed for CELSS could be used for permanent space stations on the Moon, Mars, and perhaps other bodies of the Solar System. A peculiar property of plants is of special interest in this respect. They grow very well at atmospheric barometric pressures much lower than those prevalent on Earth. When the composition of the air is carefully balanced (especially humidity, CO_2, and low oxygen content) plants may be grown at less than a tenth of the Earth's sea level pressure of 1 bar. This would much ease construction problems on places of almost zero or extremely low atmospheric pressure, such as the Moon or Mars, respectively (Richards *et al.* 2006). Present NASA plans for a human expedition to Mars call for first setting up a way station on the Moon. The way station would probably be supported by a CELSS unit (McKay 2007).

The bottom line for human space travel with present technology is that it must be limited to nearby regions of the Solar System. Even so, there is a very legitimate argument as to whether the enormous allocation of resources required to support human exploration to even so near a space target as Mars could not be better spent. The alternatives are automatic space probes or Earth-bound or Earth satellite research instruments. A further problem, whether going to Mars or elsewhere and whether piloted or automated probes are used, is how to prevent cross contamination (Rummel and Billings 2004).

Limitations to the exploration of space with robotic probes

Automated, robotic probes sent to the planets, comets, and asteroids of the Solar System have been extraordinarily successful. While providing fascinating physical and chemical data, they have indicated sites where at least primitive, Earth type life forms could possibly exist.

Life-seeking robotic probes were landed on Mars decades ago, and there have since been many others which have landed on or orbited the planet (most recently the NASA Phoenix probe in May 2008). In early 2009 two such probes are still traversing the Martian surface and another is orbiting the planet, supporting the 2008 arrival of Phoenix. So far no convincing evidence of life, extinct or extant, has been found. However, there is still hope of finding signs of life, possibly organic molecules, in subsurface layers close (cms) to the surface, which have been found to contain frozen water. In addition to Mars, other planetary satellites in the Solar System such as Jupiter's Europa or Saturn's Enceladus appear to have liquid water. Life that is independent of photosynthesis may have developed there.

Robotic probes to other solar systems are well beyond our present capabilities. In 1978 two probes, Pioneers 10 and 11 were launched and directed into orbits around the planets Jupiter and Saturn. After successfully completing their missions, they continued along trajectories which would eventually take them out of the Solar System. They are currently heading in the direction of the 'nearby' star, Aldebaran, 'only' 68 ly from the Sun. After three decades of flight they are still in the Solar System, at a distance from Earth of some 14 G km. Radio communication has only recently been lost. It will be 2 million years from now before they reach the region of Aldebaran. Pioneer 10 carries a plaque inscribed with information on Earth and humanity—a charming if whimsical touch. Assuming the Aldebarans (if they exist) promptly return us a message encoded in a narrowly directed radio or laser beam, it will be at least 2 My (+68) before it arrives. Will we still be around to receive their response?

What is the probability of there being other intelligent civilizations in the universe?

The limited regions in the universe where Earth-like environments could exist were discussed in Chapter 2. Although the probability of conditions amenable to life existing in any one place is small, the enormous size of even our galaxy, with its approximately 400 G stars, makes it difficult to accept that ours is the only planet where intelligent life has developed. This approach is the so-called 'theory of mediocrity'—we are not so exceptional. Moreover, with respect to the entire universe, the number of star systems in our galaxy with conditions amenable to life must be multiplied by the equally large number of galaxies. To carry the argument to an even greater extreme, some cosmologists, especially those dealing with string theory (the 'theory of everything' uniting relativity and quantum physics) believe that we may be living in only one of innumerable parallel universes—but the mind boggles!

Meanwhile, within the 'narrow' confines of the Milky Way there are three basic questions which have been and are being addressed:

1. Can we estimate the number of intelligent civilizations in our galaxy?
2. If there are other intelligent civilizations, why have they not contacted us?
3. Can we detect the presence of other civilizations?

The Drake equation

One of the first attempts to answer question (1) was made by the astronomer Frank Drake in 1961. There are a number of forms of this equation. One of the most familiar is shown in Box 10.1.

Drake himself, and all those after him, did not consider his equation to be an algorithm which would deliver an answer in the form of a definite number. It is more a framework for considering the parameters which must be taken into account. Progress in evaluating most parameters has been slow, but for at least one it has been very rapid. At the time Drake formulated his equation the second factor (f_p) 'the fraction of those stars that have planets' was completely unknown. Today it appears that at least 15% of the stars do have planets.

Other factors in the equation remain almost unknown and perhaps unknowable. The one which seems to be most contentious is the fraction of inhabitable planets with intelligent life (f_i). First estimates start from the example of Earth, where life seems to abound and humanity has reached a level of technology allowing for a search of the universe for other life. However, biologists, first and foremost among them the late and highly distinguished Ernst Mayr, have pointed out just how fortuitous was the appearance of intelligent, technological humans on Earth. Mayr's arguments appeared in a debate with astronomer Carl Sagan, published on the pages of the newsletter of the Planetary Society, beginning in 1995 (see Sagan/Mayr debate in References and Resources, Internet portals).

Box 10.1 The Drake equation

The Drake equation states that

$$N = R^* \times f_p \times n_e \times f_l \times f_i \times f_c \times L$$

where

N is the number of civilizations in our galaxy with which communication might be possible,
R^* is the average rate of star formation in our galaxy,
f_p is the fraction of those stars that have planets,
n_e is the average number of planets that can potentially support life per star that has planets,
f_l is the fraction of the above that actually go on to develop life at some point,
f_i is the fraction of the above that actually go on to develop intelligent life,
f_c is the fraction of civilizations that develop a technology that releases detectable signs of their existence into space,
L is the length of time for which civilizations release detectable signals into space.

Mayr emphasized the following considerations, which are somewhat expanded here:

- Life has existed on Earth for 3.7 G years, more than a quarter of the age of the universe, but there is basically just one form of life on Earth.
- Only after 1.8 G years of evolution did the first multicellular eukaryotes appear.
- Life on Earth is divided into bacteria and fungi, plants, and animals, but nervous systems (and brains) developed only in animals.
- Animals developed some 60–80 lines but only in the chordate vertebrates did intelligence appear.
- Within the vertebrates there are fish, amphibians, reptiles, birds, and mammals, but higher intelligence developed only in mammals.
- There are some 24 orders of mammals, but only one of hominids.
- The first hominids appeared some 3–6 My BP, but *Homo sapiens sapiens* evolved only in the last 200,000 years;
- *Homo sapiens sapiens* produced some 20 civilizations, but only the most recent developed technology capable of seeking out other intelligent life.
- The development of modern technology took place only once and this in the last 100 years of the 4.55 Gy lifetime of Earth.

Contrary to what many non-biologists believe, the evolution of intelligence and technological civilizations is not a *necessary* outcome of evolution. It was a purely chance event of extremely low statistical probability. The average lifetime of a higher species is about 100,000 to 1 My. Only once in biological history did higher intelligence, followed by technological ability, evolve. This mutation or mutations, could have turned up in fish, dinosaurs, or elsewhere, or may not have occurred at all. Moreover, it has yet to be shown that, from the point of view of the criteria of evolutionary success (mainly fecundity and survival), a large brain is advantageous. The likes of Shakespeare, Beethoven, Newton, or Einstein are of little value to evolution. Will we still survive 100,000 y from now? At the present time our technological advance appears to have far outstripped our sociological abilities. We are good at reproduction, but must still prove that we can survive.

Ward and Brownlee, in their already classic book *Rare Earth* (2000), are among the foremost in pointing out the enormous number of chance events, astronomical, geological, and biological, which allowed the evolution of advanced, intelligent life on Earth. Their conclusion is that while unicellular life may be quite common in the universe, intelligent life would be almost inconceivably rare. Their argument is not only biological but, as described above, includes the very small chance of the physico-chemical conditions supporting life being formed and sustained for billions of years on another planet otherwise similar to Earth.

Carl Sagan's reply to Mayr was based mainly on the vastness of the galaxy, even after taking into account that large sections would be uninhabitable. This argument continues to sustain the belief (hope? fear?) of many that we may one day be able to contact, or at least hear from, extraterrestrial civilizations (see SETI below).

The literature shows estimated solutions to the Drake equation—giving the number of intelligent, communicating civilizations in the galaxy—ranging from 0 to about 10,000. At this time (early 2009) we really cannot do any better.

The Fermi paradox

Enrico Fermi was one of the greatest physicists of the 20th century. He received the 1938 Nobel prize in physics for his work on the atomic nucleus. At the beginning of the Second World War Fermi fled to the USA from fascist Italy to save his Jewish wife and children. He was quickly brought into the Manhattan Project, in Los Alamos, New Mexico, where the first atomic bombs were developed, ushering in the nuclear age.

A possibly apocryphal story tells of a lunch break, in 1950, at which Fermi suddenly and famously blurted out 'Where is everybody?'. Fermi was not interested in the lack of attendance that day at the canteen, but rather in the above question (2) (Why have they not contacted us?). At that time there had been a spate of flying saucer stories. While not taking them seriously, Fermi began a number of quick mental calculations. He reasoned that if only one in a million stars of the Milky Way had planets with intelligent beings and they began space travel, then within a few tens of million years they should have spread out and reached all regions of the Galaxy, including our Solar System. So why have we not seen them? Why have they not made their presence known? This conundrum became known as the Fermi paradox.

According to one wit working in astrobiology: 'There are two answers to the question of whether we are alone in the universe, and each is equally terrifying: (1) We are alone, and (2) We are not alone'. Answer (1) solves the Fermi paradox.

Fermi's question was of course based on the assumption that we are not alone and that there are other intelligent entities out there. Numerous solutions have been suggested to the Fermi paradox, all in the realm of speculation. Stephen Webb (2002) collected 50 propositions, all of which are reasonable in so far as they don't break with currently accepted physical law. The one I find most attractive (a purely subjective choice, with little reason to prefer it over the other solutions) is the 'zoo theory'. There are a number of versions, but it goes approximately as follows:

- Humans are currently restricted to spaceflight within their own Solar System. This, for all the reasons enumerated above.
- Any civilization able to overcome these limitations would be vastly more advanced than ours. They may consider us and our science and technology, not to mention our social abilities, characterized by wars, famine, disease, poverty, crime, drugs, etc., as we would regard a new and dangerous species of ape, discovered in some out of the way jungle.
- If we are lucky and they don't exterminate us in the interest of sanitation in the universe, they may assign us to conservation and study, under zoo conditions.

We may be in such a zoo, surrounded by a virtual fence which prevents us from roaming and polluting the galaxy. The fence is perhaps 'painted' with the view our keepers want us to see, which does not include seeing them. Is there any proof (or disproof) of such a hypothesis? No, and probably almost by definition, none is possible.

SETI—the search for extraterrestrial intelligence

The above considerations would seem to make it very audacious for a scientist to devote much of his or her life's work to the search for intelligence outside of Earth. Yet the vast size of the universe and the enormous potential impact of a positive outcome have made a few intrepid souls venture in this direction. Central to the thinking behind this work is the theory of mediocrity, mentioned above. It is indeed difficult to accept that in a cosmos so vast humanity is not a normal (even if rare and random) outcome of physics, chemistry, and biology.

At this stage the search is for signals from extraterrestrials. Two-way communication with an extraterrestrial intelligence is not considered possible with present science. The distances involved and the time required for EM transmissions are just too great for our limited resources and short lives.

Extraterrestrials with highly sensitive radio antennae may be able to learn of us from our radio and television transmissions. Note that among the first TV broadcasts was Hitler opening the 1936 Olympic Games. Shortly afterwards came the Nazis, marching into Vienna and Paris; hardly an attractive introduction for humanity. This may be a PR concern, but only in the future. Even a 'nearby' civilization, just 100 ly from Earth, will only be picking up these broadcasts some 30 years from now.

We are not currently able to receive such local broadcasts, intelligence-generated EM radiations, from extra-Solar System planets. The incoming signal would be too weak for the largest of our antennae, with the greatest possible amplification. Consequently, SETI efforts have been concentrated on the search for signals or signatures which are broadcast with the intention of making their presence known to residents of other star systems. Such signals may have been sent out hundreds of years ago. There may be another problem. The sending of such signals may be preceded by a search for planets having inhabitants with technologies capable of receiving EM radiation communications. If Earth had been surveyed, say in 1800, it would have been concluded that we are a non-technological civilization. So the extraterrestrials would not have bothered to direct their signals in our direction. This is something of a logical impasse, for now and the next few hundred years.

Our local galaxy, the Milky Way, contains about 400 G stars. The first steps of SETI must be limited to just a few promising star systems, which may have the potential to support life (Chapter 2). It is also necessary to select wavelengths of the EM spectrum surveyed, from among the very many possible. Finally, there is the problem of criteria for recognizing an intelligent signal from amongst the electronic noise of the galaxy. How will the extraterrestrials modulate their signal? What code would they use? Scepticism aside, and to briefly address the last question first, narrow band-width signals, of steady if modulated intensity, with a fixed pattern, whose strength exceeds background radiation, would be of obvious interest.

A seminal paper on SETI was that by Cocconi and Morrison (1959). They analysed the different EM bands from the point of view of interstellar communication. They concluded that radiation at frequencies below 1 GHz would be masked by galactic 'noise'. Above 10 GHz oxygen and water molecules in our atmosphere would absorb and conceal incoming signals. The lower end of this range is preferred as the higher the frequency the greater the Doppler shift interference resulting from planetary motion. Of special interest is the 1.42 GHz frequency at which hydrogen emits. It is commonly studied as a basic indicator frequency in radioastronomy. Consequently, any signal sent at this frequency and intentionally modulated, in a repetitive pattern, would have a relatively good chance of being detected.

In 1960 Frank Drake, of Drake equation fame, commenced one of the first searches for extraterrestrial intelligence, at Cornell University. At about the same time a number of SETI projects were begun in the Soviet Union.

Since the early SETI efforts there have been numerous other projects, with more and more radio wavelengths and source target stars. A major problem is the analysis of the vast amounts of data which accrue. This was partly solved by parallel computing, in which the general public was invited to participate using the spare time on their home computers. Participants were sent an initial analysis program (SETI@home) and sets of data over the internet. The data sets were analysed by desktop computers (Fig. 10.4) and the results returned to Berkeley University, California, for further analysis. This turned out to be the greatest cooperative research project in the history of science, with about 5 million participants!

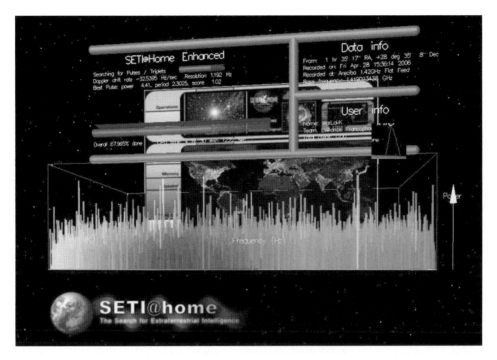

Fig. 10.4 A screen from the SETI@home distributed parallel computing program, used to crunch SETI data on some 5 million home computers. Courtesy of the SETI institute.

If an alien civilization is broadcasting in all directions, the transmission received by Earth, from a source only 50 ly away, would be extremely weak. This would be so unless the source intensity far exceeds the level that we on Earth would be able to generate with our present energy resources. Alternatively, the signals may be sent in a narrow beam of high intensity.

Such narrow, high-intensity beams are best generated by lasers, currently (by today's technology) mainly in the optical wavebands. There are two inherent problems for interstellar communication. First, laser beams are monochromatic; meaning that they are of one frequency only. This raises the question of which frequency to look for? The second is that being of very narrow width, there is little chance of such a beam being intercepted, unless it is intentionally sent in the direction of the target. Moreover, such laser signals would be far more likely to be absorbed by interstellar dust. Even so, there are a number of optical SETI projects currently in progress (2009), including one at Berkeley and another (with the Smithsonian Institute) at Harvard, looking for such laser beam signals.

Following on from some earlier suggestions, Christopher Rose and Gregory Wright (2004) calculated that sending an inscribed physical object may be a much more efficient method of dissemination of information than EM communication. This holds true especially for one-way communication. In this argument 'efficient' means the amount of energy that must be expended to send a unit of information. This is the old 'message in a bottle' approach, as implemented in the 1978 Pioneer probe targeted toward Aldebaran, mentioned above. According to this thesis an extraterrestrial may have already sent us an artefact with a DVD or similar or more advanced form of physical data compression and storage. Such an artefact would most reasonably have been 'parked' in one of the Earth–Sun or Earth–Moon Lagrangian positions,[1] awaiting its detection by an emerging technological civilization. Some efforts have already been made to search for such artefacts in the Lagrangian planes.

The sad bottom line for this, and all other SETI efforts over the last half century, is that no signal indicating the existence of an extraterrestrial intelligence has been received. SETI enthusiasts point out that we have still surveyed only a small fraction of the potentially promising targets. If, however, the search remains unfruitful we are returned to the Fermi paradox or to the alternative possibility, that we are indeed alone.

Human spaceflight to extra-solar star systems

With the human life span less being than 120 years, and a useful astronaut working life of perhaps 20 or 30 y, piloted flights to extra-solar star systems within the Milky Way just do not seem possible. Even science fiction rarely tackles travel to other galaxies. Andromeda, the nearest galaxy after Sagittarius (which by some estimates is within the Milky Way) is at a distance of 2.2 M ly. Science fiction scenarios in which humanity flees a depleted Earth to colonize planets of other star systems is consequently beyond all reasonable present-day forecasts. This pessimistic view holds even after taking into account the most optimistic projections for future developments in science, technology, and medicine.

In the past, predictions such as the ultimate speed of aeroplanes being limited to that of sound in air, or the genetic code being too large to be unravelled, have often been swept aside by unexpected scientific and technological breakthroughs. In the 3rd century CE the authors of the Talmud wrote that 'today only fools and the very young prophesy'. That maxim still holds true in this 21st century. So perhaps the predictions of science fiction will one day be realized, as they have so often in the past. Submarines, heavier-than-air flying machines, supersonic flight, travel to the Moon, antibiotics, communication satellites, world-ranging pocket phones, computers, the internet, genetic engineering, etc., etc., are but a few examples of this perspicacity. All appeared in literature long before they were invented, or thought possible. They are, after all, an expression of the wonderful human ability to extrapolate, imagine, and dream.

---

[1] Regions in the orbital planes of two bodies, revolving around each other, such as the Earth and Moon, where a third body, of relatively negligible mass, would be held in gravitational equilibrium.

Summary

- The question of whether there is life elsewhere in the universe, other than on Earth, is at least as old as recorded history. It was discussed by the Babylonians, Chinese, and Indians as long ago as 2000 BCE, and by the Greeks and Romans from about 1000 BCE to 400 CE. The arguments continued through the Enlightenment, to this very day. Today, the generally accepted opinion is that primitive life may exist in some places in the Solar System and elsewhere in the universe, but that there is no intelligent extraterrestrial life in the Solar System. It is probably very rare or non-existent elsewhere.
- Robotic probes are being used to survey the Solar System, with great efficacy. Beyond the Solar System time and distance factors make them impractical.
- Piloted spaceships have reached the Moon and flights to Mars are being planned. Longer flights do not seem possible in the coming decades. Spaceflight outside the Solar System does not seem possible with our present-day science and technology. Space exploration by humans is constrained by our short life span, fragility, slow travel speeds, the vast cosmic distances, and cost.
- Recent findings of numerous planets orbiting other stars have encouraged the search for signatures of extra-Solar System life. This search is necessarily confined to observations and analysis of radiations reaching Earth (mainly EM, but some particulate).
- The number of other potentially communicating civilizations, from among the ~400 G star systems in the Milky Way, is impossible to calculate. This is because we know with certainty of only a single incidence—on Earth. Estimates vary between 0 and 10,000.
- As of today (early 2009) no unequivocal evidence has been obtained for the existence of extraterrestrial life within or outside of the Solar System whether primitive or advanced; but the search continues.
- The study of the astrobiology of Earth indicates that our biosphere was enabled to appear, evolve, and persist for 3.7 Gy as a result of a rare combination of dynamic cosmic and geological factors. Earth's environment is unstable and easily upset. Biosphere perturbations may be caused by changing geological and cosmic forces, and also by biological factors, including those resulting from human activity.
- Human colonization of other star systems does not seem possible at this time. For the foreseeable future, Earth will be our only home.

A short epilogue to the *Astrobiology of Earth*

This is the way the world ends
This is the way the world ends
This is the way the world ends
Not with a bang but a whimper.

T. S. Eliot, The Hollow Men [written in 1925,
before the nuclear age]

Whichever way we look at our past, present, and foreseeable future, the take-home lesson is clear. Only a chance combination of dynamic cosmic and geological factors allowed life, in the currently human-favourable biosphere, to appear, evolve, and persist on Earth. These many factors have in the past determined and affected the physical–chemical–biological environment of the biosphere , and continue to do so today. This environment is not stable but is in a state of quasi-equilibrium. Apart from the effects of uncontrollable geological and extraterrestrial forces, the biosphere is being perturbed in an unplanned, chaotic manner by us humans. The outcome can only be guessed at. It could be catastrophic—if not for the pliable biosphere, then at least for the human species.

Terra forming of other planets of our Solar system, such as Mars, to make their environments amenable for humans, is most unlikely. Furthermore, flight to planets on other star systems is not possible with our present science and technology. For now, Earth is our only home. We must conduct ourselves wisely and learn to cherish and preserve it; otherwise a nuclear bang or perhaps just the 'whimper' may be our destiny.

Appendix–A
A workshop guide

The topic of astrobiology of Earth draws from astronomy, geology, general biology, evolution, ecology, and many other disciplines. If presented by multiple specialists, standards will be high and the material presented very authoritative. But this could result in a pastiche of disjointed presentations, with an unclear central theme. A solution to this problem in didactics is having a workshop format, with a single class leader or lecturer. This puts considerable onus on the workshop instructor, which can be partly alleviated by first inviting a few outside lecturers to introduce their particular subjects, from the point of view of astrobiology. Students then choose topics from among set subjects as, for example given here chapter-by-chapter, or in the different chapter sections. Their papers could be based upon the expansion and update of this material. The book text is at a level that enables the participation of students from many different university departments. The References present many papers, reviews, and texts for further reading and for preparation of class papers.

The following suggestions for course organization are based on experience obtained over the last 8 years. Students of our Astrobiology of Earth workshop come from general biology, astronomy, ecology, and earth sciences. A few also come from medicine and the humanities. They take the course as an elective in the final year of their BSc studies, or as one of their post-graduate courses.

The workshop is slated for a nominal 3 hours a week, during a single semester of some 14–15 weeks. In practice the class meets for one 2-hour period a week. The students are credited for a third hour, taking into consideration a heavy home assignment and at least two personal tutorials. The format will of course be modified according to particular local needs and circumstances.

The organization of the workshop is explained briefly at the first meeting, where students also receive a 2-hour introductory lecture that covers most aspects of the course. The aim here is to set out the central theme, putting all the topics into a single perspective. It also demonstrates that the entire subject, and consequently each of the specific topics, can be presented in the space (written paper) and time (class presentation) allotted. This limitation almost invariably raises doubts! It is also emphasized that there is no harm in overlap between subjects as this aids learning and, where differences in interpretation and approach are revealed, this is all to the good.

At the end of the introductory talk students choose their individual subjects from a list of some 20 topics. These have varied over the years, and will no doubt be modified according to the lead lecturer's special interests. After the introductory lecture, students receive short presentations from experts in the fields of astronomy, geology, origin of life, and evolution. These take place over the next 4 or 5 weeks. Talks given by the invited lecturers often overlap the student presentations but, as noted above, this is not considered detrimental.

Each chapter gives a brief summary and references to the recommended reading/research list. Unlike regular textbooks, the material in these chapters is only a guideline covering the central concepts. Students are encouraged to introduce other, and preferably their own original viewpoints, while adhering to the main workshop theme, as set out in the Introduction.

Each student, or at most two, prepares a paper and class presentation. There is some pressure on the time available for preparing the first papers, which is partly relieved by having the invited lectures in the first weeks of the course. A minimum of 1 month is allowed for preparation of the first papers.

An invited astronomer is requested to put special emphasis on quantification in the cosmos, a very difficult subject for the non-astronomer. This includes orders of magnitude, numbers, distances, mass, speed and time, and factors on the astronomical scale which may predispose or prohibit life. A geologist is asked to explain how the physical/chemical factors of the Earth varied over time and affected the biosphere, with special emphasis on techniques used for dating strata. A biologist reviews present ideas on the origin and genesis of life. An evolutionist is invited to talk on the various theories of the mechanisms of Darwinian evolution.

Each student or pair of students meets with the lecturer for two tutorials. In the first they are guided as to what they are expected to produce, and how. The first assignment is a term paper of some six to nine pages, setting out the subject within the framework of the workshop. For example, if the subject is 'The Sun in the Galaxy' students must explain the significance to life of the Sun's situation, and mineral composition and the biological importance of its radiations. The emphasis is always on the importance to life on Earth. Special emphasis is put on the clear and uncluttered preparation of the papers that become a main study tool for the other participants. Students are expected to use the sourcebook list to prepare their papers. However, books and reviews are inevitably a few years behind the latest published research results, even as they go to print. Consequently, students are instructed to supplement their research with papers derived mainly from databases of published, peer-reviewed, journal papers of recent years. When using the internet it is important for them to recognize the advantages and pitfalls of this unedited data source. A draft of the student paper is discussed at a second tutorial, before its final approval—see below.

The second student assignment is the preparation of a 30-minute (net) 45-minute (gross) class presentation. The latter is not so much a formal seminar as a workshop discussion. About a week before the presentation, the corrected written paper and the presentation are posted on the class internet homepage. Students are expected to review this material and come prepared. Interrupting the presentation with comments, questions, and discussion, by students and the workshop leader, is not only allowed but encouraged. Microsoft Word is used for the paper and Power Point for the class presentations (no commercial for Mr Gates intended, but he seems to have cornered the market!). Where necessary, students are also given some guidance on the techniques and tricks of class lecturing. Three points which it has often be found necessary to emphasize are: the need to avoid overloading the screen with innumerable curves, numbers, and notes; never reading, whether from notes or from the screen; and not overdoing Power Point's entertaining, but sometimes distracting, 'bells and whistles'.

Examination and class grades

At the end of the workshop students are examined on all the material of the course. The 2-hour exam takes the form of some 20 multiple-choice questions, followed by two half-page essays chosen from some five or six questions. Half the exam grade is given for the multiple-choice questions and half for the essay questions. The multiple-choice questions include some designed to test the appreciation of quantity: number, size, degree, and time. For example, students are expected to have at least an approximate idea of the age of the Earth, the timing of the Cambrian explosion, the original and present oxygen content of the atmosphere, the number of stars in our galaxy, the solar radiation load at the top of and under the atmosphere, etc., etc. The essay exams are intended to test originality and integrative thinking and to gauge whether any benefit has been obtained from at least some of the recommended reading list. The final workshop-course grade is derived from the term paper (30%), presentation (20%), and the final examination (50%). The final grade is adjusted to take into consideration active class participation.

We hope that these workshops will open new intellectual vistas and invite many questions. Students continuing to graduate-level studies are encouraged to search for answers to their in-depth Astrobiology of Earth questions within the framework of research thesis projects. These are carried out in specialist departments, such as Earth Sciences or Astronomy.

Appendix–B
Notes added 'in press'

Astrobiology and within it Astrobiology of Planet Earth, is such an eclectic subject that almost all new findings of science, technology and even the economy are relevant. The following are just a few, recent, almost random items of importance to the overall theme of this book.

– Clouds of Carbonaceous aerosols, from the burning of wood, dung and coal, have become so widespread in Asia as to increase the temperature of the Earth's atmosphere, while cooling its surface. This affects the calculations of the assumed anthropogenic climate change, thought to raise surface temperatures (Gustafsson et al *Science* (2009) 23:495–498).

– Analysis of the growth of stalactites (related to precipitation) in a cave in China, over a time scale of the last 2000 years, has revealed periods of extreme, prolonged drought in the 9th century. This correlates with the end of the Ting dynasty in China and the Mayan civilisation in Central America. It brings further evidence of large, world scale climate change in historical times, whose cause has been unrelated to human activity. (Zhang et al, *Science* (2008) 322: 940–942).

– Worldwide degradation of corals has been said to be a result of a rise in sea temperature, part of the putative anthropogenic climate change. Rosenburg et al, in *Nature, Reviews of Microbiology*, March 26, 2007, report that micro-organism infections may be implicated. Although this requires further worldwide corroboration, it seems a more reasonable hypothesis than sea water temperature rise, which to date, has been < 1°C, in the last century. (This is the figure for air temperature; the change in sea temperature is even less).

– In discussing world energy and food production it was stated that oil had reached $145 a barrel and that introduction of ethanol and bio-diesel, while being of doubtful utility, have been falsely accused of being entirely responsible for the increase of world food prices. The world economic recession in the second half of 2008 has changed these calculations. As this goes to print, in early 2009, the price of oil has fallen to < $40 a barrel. This makes alternative energy sources, such as ethanol and solar voltaic, even less attractive. Of greater importance to the poor of the world, food prices have dropped, as energy costs for production and transport have gone down. However, even the most pessimistic economists predict an end to the crises after 2011. This will again place the problem of providing energy and food for the ~9 G population, predicted for the year 2050 (up from 6 G, at present) in the forefront of human concern.

– Another disadvantage has recently been pointed out in relation to the production of fuel from agricultural crops. Bio-fuel production presents a huge demand for fresh water. 4,000 liters of water are required to produce 1 liter of ethanol and 9,000 for diesel fuel. This in a world in which, by 2025, one third of the population will be desperately short of irrigation, industrial and drinking water (Brabeck-Letmathe, writing in the October 5, 2008 edition of the *International Herald Tribune*).

– New "adaptive optics" has, for the first time, enabled direct observation of planets around another star – Fomalhaut, at a distance of 25 ly. This new telescope technology overcomes the glare of the parent star, which previously limited the detection of extra-solar system planets to indirect methods. (see *Science*, November 14, 2008). A new planet detector, launched into space orbit, in early 2009, is expected to reveal far more details of the newly detected extra solar system planets.

– The text discussion of Dark Matter states that, although it is thought to have five times the mass of normal matter (in the universe) it has no known affect on life. Life and the planets on which it may be found, are made from normal matter. Adler, in the October, 17, 2008 edition of the *Journal of Physics*, calculates that the presence of Dark Matter around the planets of our Solar System (including Earth) may be responsible for at least part of their internal warming. Moreover, it may be responsible for the navigational anomalies found in the orbits of probes sent from Earth to the Solar System.

– An important paper, offering a solution to the conundrum of Earth's almost stable temperatures, over millions of years, was inadvertently omitted from the Reference list. It is: Walker et al, (2003) A negative feedback system for the long term stabilization of Earth's surface temperature. *Journal of Geophysical Research*, 86: 9776–9782.

References and Resources

References

Ambrose, S.H. (1998). Late Pleistocene human population bottlenecks, volcanic winter, and differentiation of modern humans. *Journal of Human Evolution*, 34: 623–651.

Andrew, C.S. and Glasspool, I.J. (2006). The diversification of Paleozoic fire systems and fluctuations in atmospheric oxygen concentration. *Proceedings of the National Academy of Sciences USA*, **103**: 10861–10865.

Anonymous (2007). *The limits of organic life in planetary systems*. Report of the U.S. National Research Council (see http://www.nap.edu/catalog.php?record_id=11919).

Arrhenius, S. (1908). *Worlds in the making*. Harper, London.

Baaske, P., Weinert, F.M., Duhr, S., Lemke, K.H., Russell, M.J., and Braun, D. (2007). Extreme accumulation of nucleotides in simulated hydrothermal pore systems. *Proceedings of the National Academy of Sciences USA*, **104**: 9346–9351.

Bada, J.F. and Lazcano, A. (2002). Some like it hot, but not the first biomolecules. *Science*, **296**: 1982–1983.

Baker, J. (2006). Tiger, tiger, burning bright. *Science*, **311**: 1388.

Barrow, J.D. and Tipler, F.J. (1996). *The anthropic cosmological principle*. Oxford University Press, Oxford [reissue of 1986 edition].

Battey, N.H.(2000). Aspects of seasonality. *Journal of Experimental Botany*, **51**: 1769–1780.

Beatty, J.K., Collins Petersen, C., and Chaikin, A.(eds) (1999) *The new Solar System*. Cambridge University Press, Cambridge.

Beatty, J.T., Overmann, J. Lince, Michael T., Manske A.K., Lang A.S., Robert E. Blankenship,R.E., Van Dover,C.L., Tracey A. Martinson, T.A., and Plumley, F.G.(2005). An obligately photosynthetic bacterial anaerobe from a deep-sea hydrothermal vent. *Proceedings of the National Academy of Sciences USA*, **102**: 9306–9310.

Beerling, D. (2007). *The emerald planet*. Oxford University Press, New York.

Beerling D.J. and Berner R.A. (2005). Feedbacks and the coevolution of plants and atmospheric CO_2. *Proceedings of the National Academy of Sciences USA*, **102**: 1302–1305.

Bell, J. (2006). The Red Planet's watery past. *Scientific American*, **295**: 40–47.

Benestad, R.E. (2002). *Solar activity and Earth's climate*. Springer Praxis, Berlin.

Bennett, J., Shostak, S., and Jakosky, B. (2003). *Life in the universe*. Addison Wesley, San Francisco.

Benton, M.J. (2003). *When life nearly died*. Thames and Hudson, London.

Bergman, N., Lenton, M.T.M., and Watson, A.J.(2004). COPSE: a new model of biogeochemical cycling over Phanerozoic time. *American Journal of Science*, **304**: 397–437.

Berner, R.A. (1997). The rise of plants and their effect on weathering and atmospheric carbon dioxide. *Science*, **276**: 544–546.

Berner, R.A. (1999). Atmospheric oxygen over Phanerozoic time. *Proceedings of the National Academy of Sciences USA*, **96**: 10955–10957.

Bindeman, I.N. (2006). The secrets of supervolcanoes. *Scientific American*, **296**: 27–33.

Brack, A. (ed.) (1998). *The molecular origins of life*. Cambridge University Press, Cambridge.

Brenchley, P.J. and Harper, D.A.T. (1998). *Paleoecology: ecosystems, environments and evolution*. Chapman and Hall, London.

Broecker, W.S. (1988). *How to build a habitable planet.* Eldigio Press, New York.

Brooks, D.J., Fresco, J.R., Lesk, A.M., and Singh M. (2002). Evolution of amino acid frequencies in proteins over deep time: inferred order of introduction of amino acids into the genetic code. *Molecular Biology and Evolution,* **19**: 1645–1655.

von Caemmerer, S. and Furbank, R.T. (2003). The C-4 pathway: an efficient CO2 pump. Photosynthesis Research, 77: 191–207.

Caldeira, K. and Rampino, M.R. (1991).The mid-Cretaceous superplume, carbon dioxide and global warming. *Geophysical Research Letters,* **18**: 987–990.

Carroll, S.B., Prud'homme, B., and Gompel, N. (2008). Regulating evolution. *Scientific American,* **298**: 38–45.

Chaloner, W. (2003). The role of carbon dioxide in plant evolution. In: Rothschild, L.J. and Lister, A.M. (eds) *Evolution on planet Earth,* Ch. 5. Academic Press, New York.

Chen, I.A. (2006). The emergence of cells during the evolution of life. *Science,* **314**: 1558–1559.

Chi, A., Valenzuela, L., Beard, S., Mackey, A.J., Shabanowitz, J., Hunt, D.F., and Jerez, C.A. (2007). Periplasmic proteins of the extremophile *Acidithiobacillus ferrooxidans.* A high throughput proteomics analysis. *Molecular and Cellular Proteomics,* **6**: 2239–2251.

Chyba, C.F. and Phillips, C.B. (2001). Possible ecosystems and the search for life on Europa. *Proceedings of the National Academy of Sciences USA,* **98**: 801–804.

Clancy, P., Brack, A., and Horneck, G.(2006). *Looking for life, searching the Solar System.* Cambridge University Press, Cambridge.

Cleaves, H.J. II and Chalmers, J.H. (2004). Extremophiles may be irrelevant to the origin of life. *Astrobiology,* **4**: 1–9.

Cleland C.E. and Chyba C.F. (2002). Defining 'life'. *Origins of Life and Evolution of the Biosphere,* **32**: 387–393.

Cleland, C.E. and Copley, S.D. (2005). The possibility of alternative microbial life on Earth. *International Journal of Astrobiology,* **4**: 165–173.

Cocconi, G. and Morrison, P. (1959). Searching for interstellar communication. *Nature,* **184**: 844–846.

Cockell, C.S. (1999). Life on Venus. *Planetary and Space Science,* **47**: 1487–1501.

Cockell, C.S. (2002). Photobiological uncertainties in the Archaean and post Archaean world. *International Journal of Astrobiology,* **1**: 31–38.

Cockell, C.S. (2005). Habitability. In: Horneck, G. and Rettberg, P. (eds) *Complete course in astrobiology,* Ch. 6. Wiley-VCH, Weinheim.

Cohen, D. (1967). Optimizing reproduction in a randomly varying environment when a correlation may exist between the conditions at the time of choice has to be made and the subsequent outcome. *Journal of Theoretical Biology,* **16**: 1–14.

Cohen, D. and Mangel, M. (1999). Investing for survival of rare severe stresses in heterogeneous environments. *Evolutionary Ecology Research,* **1**: 987–1002.

Cohen, J.E. (1997). Population, economics, environment and culture: an introduction to human carrying capacity. *Journal of Applied Ecology,* **34**: 1325–1333.

Cohen, Y.E. (2005). Human population grows up. *Scientific American,* **293**: 26–33.

Comins, N.M.(1993). *What if the moon didn't exist.* Harper Collins, New York.

Conway, C. (1997). *The doubly green revolution: food for all in the twenty first century.* Cornell University Press, Ithaca, NY.

Courtillot, V. (1999). *Evolutionary catastrophes—the science of mass extinction.* Cambridge University Press, Cambridge.

Crutchfield, J.P. and Schuster, P. (ed.) (2003). *Evolutionary dynamics. Exploring the interplay of selection, accident, neutrality and function.* Oxford University Press, Oxford.

Darling, D. (2001). *Life everywhere.* Basic Books, New York.

Darwin, C. (reprint of 1859 original, 2003). *The origin of species.* Signet Classics, New York.

Davies, P. (2000). *The fifth miracle. The search for the origin and meaning of life.* Simon and Schuster, New York.

Davies, P. (2007). Are aliens among us? *Scientific American*, **297**: 36–43.

Davies, P. and Lineweaver, C. (2005). Finding a second sample of life on Earth. *Astrobiology*, **5**: 154–163.

Dawkins, R. (1986). *The blind watchmaker.* Penguin Books, London.

Dawkins, R. (1991). *The selfish gene.* Oxford University Press, Oxford.

Delsemme, A. (1998). *Our cosmic origins.* Cambridge University Press, Cambridge.

Delsemme, A. (2000). Cometary origin of the biosphere. *Icarus*, **146**: 313–325.

Delsemme, A. (2001). An argument for the cometary origin of the biosphere. *American Scientist*, **89**: 432–442.

Dergachev, V.A., Dmitriev, P.B., Raspopov, O.M., and Jungner, H. (2006). Cosmic ray flux variations, modulated by the solar and Earth's magnetic fields, and climate changes. *Geomagnetism and Aeronomy*, **46**: 118–128.

Des Marais, D.J. (2000). When did photosynthesis emerge on Earth? *Science*, **289**: 1703–1705.

Diamond, J. (2004). *How societies chose to fail or succeed.* Viking, New York.

DiChristina, M. (ed.) (2005). Our ever changing Earth. *Scientific American*, **15**: 2.

Dick, S.J (1984). *Plurality of worlds.* Cambridge University Press, Cambridge.

Dick, S.J. (2001). *Life on other worlds. The 20th century extraterrestrial life debate.* Cambridge University Press, Cambridge.

Dick, S.J. (2006). NASA and the search for life in the universe. *Endeavour*, **30**: 71–75.

Di Giulio, M. (2003). The universal ancestor was a thermophile or a hyperthermophile. Tests and further evidence. *Journal of Theoretical Biology*, **221**: 425–436.

Ditlevson, P.D. (2005). A climatic thermostat making Earth habitable. *International Journal of Astrobiology*, **4:** 3–7.

Dornelas, M., Connolly, S.R., and Hughes, T.P. (2006). Coral reef diversity refutes the neutral theory of biodiversity. *Nature*, **440**: 80–82.

Dudley, R. (1998). Atmospheric oxygen, giant Paleozoic insects and the evolution of aerial locomotor performance. *Journal of Experimental Biology*, **201**: 1043–1050.

de Duve, C. (2005). The onset of selection. *Nature*, **433**: 581–582.

de Duve, C. (1991). *Blueprint for a cell.* Niel Patterson, Burlington, NC.

Dyson, F. (1999). *Origins of life.* Cambridge University Press, Cambridge.

Ehleringer, J.R., Cerling, T.E., and Dearing, M. (2005). *A history of atmospheric CO_2 and its effects on plants, animals and ecosystems.* Springer, Berlin.

Eigen, M. (1994). On the origin of biological information. *Biophysical Chemistry*, **50**: 1.

Eldredge, N. and Gould, S.J. (1972). Punctuated equilibria: an alternative to phyletic gradualism. In: T.J.M. Schopf (ed.), *Models in paleobiology*, pp. 82–115. Freeman Cooper and Co., San Francisco.

Engel, G.S., Calhoun, T.S., Read, E.L., Ahn, T-K., Mancal, T., Cheng, Y-C., Blankenship, R.E., and Fleming, G.R. (2007). Evidence for wavelike energy transfer through quantum coherence in photosynthetic systems. *Nature*, **446**: 782–786.

Erwin, D.H. (2006). *Extinction.* Princeton University Press, Princeton, NJ.

Evans, L.T. (1998). *Feeding the ten billion.* Cambridge University Press, Cambridge.

Evans, M.E.K., Ferrie're, R., Kane, M.J., and Venable, D.L. (2007). Bet hedging via seed banking in desert evening primroses (Oenothera, Onagraceae): demographic evidence from natural populations. *The American Naturalist*, **169**: 184–194.

Fedoroff, V. and Cohen, J.E. (1999). Plants and population: is there time? *Proceedings of the National Academy of Sciences USA*, **96**: 5903–5907.

Fenchel, T. (2002). *Origin and early evolution of life*. Oxford University Press, Oxford.

Fleminger, G., Yaron, T., Eisenstein, M., and Bar-Nun, A. (2005). The structure and synthetic capabilities of a catalytic peptide formed by substrate-directed mechanism—implications to prebiotic catalysis. *Origins of Life and Evolution of Biospheres*, **35**: 369–382.

Flindt, R. (2006). *Amazing numbers in biology*. Springer Verlag, Berlin.

Forster, C.A. (2003). Drifting continents and life on Earth. In: L.J. Rothschild and A.M. Lister (eds), *Evolution on planet Earth*, Ch. 15. Academic Press, New York.

Fortey, R.A., Briggs, D.E.G., and Wills, M.A. (1997). The Cambrian evolutionary 'explosion' recalibrated. *BioEssays*, **19**: 429–434.

Franck, S., Block, A., von Bloh, W., Bounama, C., Schellnhuber, H.-J., and Svirezhev, Y. (2000). Habitable zone for Earth-like planets in the solar system. *Planetary and Space Science*, **48**: 1099–1105.

Franck, S., von Blough, W., Bounama, C., Steffen, M., Schonberner, D., and Schellnhuber, H.-S. (2002). Habitable zones in extrasolar planetary systems. In: G. Horneck and C. Baumstark-Khan (ed.), *Astrobiology: the quest for the conditions of life*, pp. 47–56. Springer Verlag, Berlin.

Freeland, S.J., Wu, T., and Keulmann, N. (2003). The case for an error minimizing standard genetic code. *Origins of Life and Evolution of Biospheres*, **33**: 457–477.

Fry, I. (2000). *The emergence of life on Earth*. Rutgers University Press, Piscataway, NJ.

Gaidos, E., Deschenes, B., Dundon, L., Fagan, K., McNaughton, C., Menviel-Hessler, L., Moskovitz, N., and Workman, M. (2005). Beyond the principle of plenitude: a review of terrestrial planet habitability. *Astrobiology*, **5**: 100–126.

Gale, J., Rachmilevitch, S., Reuveni, J., and Volokita, M. (2001). The high oxygen atmosphere toward the end-Cretaceous; a possible contributing factor to the K/T boundary extinctions and to the emergence of C4 species. *Journal of Experimental Botany*, **52**: 801–809.

Gasperini, L., Bonatti, E., and Longo, G. (2008). The Tunguska mystery. *Scientific American*, **298**: 56–61.

Gates, D.M. (1993). *Climate change*. Sinauer, Sunderland, MA.

Gaudi, B.S. *et al.* (2008). Discovery of a Jupiter/Saturn analog with gravitational microlensing. *Science*, **319**: 927–930.

Gibbs, W.W. (2004). On the termination of species. *Scientific American*, **285**: 40–49.

Gibson, D.G. *et al.* (2008). Complete chemical synthesis, assembly, and cloning of a *Mycoplasma genitalium* genome. *Science*, **319**: 1215–1220.

Gies, D.R. and Helsel, J.W. (2005). Ice age epochs and the sun's path through the galaxy. *The Astrophysical Journal*, **626**: 844–848.

Gilmour, I. and Sephton, M.A. (eds) (2004). *An introduction to astrobiology*. Cambridge University Press, Cambridge.

Ginzburg, M. and Ginzburg, B.Z. (1985). Ion and glycerol concentrations in 12 isolates of *Dunaliella*. *Journal of Experimental Botany*, **36**: 1064–1074.

Gold, T. (1999). *The deep hot biosphere*. Springer Verlag. New York.

Goldblatt, C., Lenton, T.M., and Watson, A.J. (2006). Bistability of atmospheric oxygen and the great oxidation. *Nature*, **443**: 683–686.

Goldsmith, D. and Owen, T.(2002). *The search for life in the universe*, 3rd edn. University Science Books, Sausalito, CA.

Gonzalez, G., Brownlee, D., and Ward, P.D. (2001a). Refuges for life in a hostile universe. *Scientific American*, **285**: 60–68.

Gonzalez, G., Brownlee, D., and Ward, P.D. (2001b). The galactic habitable zone: galactic chemical evolution. *Icarus*, **152**: 185–200.

Gould, S.J. (1990). *Wonderful life*. W.W. Norton, New York.

Gould, S.J. (2002). *The structure of evolutionary theory*. Belknap, Harvard, Cambridge, MA.

Grady, M.M. (2001). *Astrobiology*. Smithsonian Institution Press, Washington, DC.

Greadel, T.E. and Crutzen, P.J. (1993). *Atmospheric change*. W.H. Freeman, New York.

Gross, M. (1997). *Life on the edge*. Perseus Books, New York.

Haddad, D., Schaupp, F., Brandt, R., Manz, G., Menzel, R., and Haase, A. (2004). NMR imaging of the honeybee brain. *Journal of Insect Science*, **4**: 7–14.

Hallam, T. (2004). *Catastrophes and lesser calamities*. Oxford University Press, Oxford.

Hallam, T. and Wignall, P.B. (1997). *Mass extinctions and their aftermath*. Oxford University Press, Oxford.

Hand, K.P. and Chyba, C.F.(2007). Empirical constraints on the salinity of the European ocean and implications for a thin ice shell. *Icarus*, **189**: 424–438.

Harrison, R.M. (ed). (1999). *Understanding our environment. An introduction to environmental chemistry and pollution*, 3rd edn. Royal Society of Chemistry, London.

Hawks, J., Hunley, K., Lee, S-H., and Wolpoff, M. (2000). Population bottlenecks and Pleistocene human evolution. *Molecular Biology and Evolution*, **17**: 2–22.

Hazen, R.M. (2005). *Gen.e.sis: the scientific quest for life's origin*. Joseph Henry Press, Washington, DC.

Hill, J., Nelson, E., Tilman, D., Polasky, S., and Tiffany, D. (2006). Environmental, economic and energetic costs and benefits of biodiesel and ethanol biofuels. *Proceedings of the National Academy of Sciences USA*, **103**: 11206–11210.

Hodell, D.A., Brenner, M., Curtis, J.H., and Guilderson, T. (2001). Solar forcing of drought frequency in the Maya lowlands. *Science*, **292**: 1367–1370.

Hoehler, T.M., Amend, J.P., and Shock, E.L. (2007). A 'follow the energy' approach for astrobiology. *Astrobiology*, **7**: 819–823.

Horneck, G. (2003). Could life travel across interplanetary space? Panspermia revisited. In: L.J. Rothschild and A.M. Lister (ed.), *Evolution on planet Earth*, p. 109. Academic Press, New York.

Horneck, G. and Baumstark-Khan, C. (ed.) (2002). *Astrobiology: the quest for the conditions of life*. Springer Verlag, Berlin.

Horneck, G. and Rettberg, P. (2005). *Complete course in astrobiology*. Wiley-VCH, Weinheim.

Horneck, G., Stoffler, D., Ott, S., Hornemann, U., Cockell, C.S., Moeller, R., Meyer, C., De Vera, J.P., Fritz, J., Schade, S., and Artemieva, N.A. (2008). Microbial rock inhabitants survive hyperve-locity impacts on Mars-like host planets: first phase of lithopanspermia experimentally tested. *Astrobiology*, **8**: 17–44.

Hoyle, F. and Wickramasinghe, N.C. (1979). *Diseases from Space*. London, Dent.

Hoyle, F. and Wickramasinghe, N.C. (1999). *Astronomical origins of life. Steps towards panspermia*. Kluwer Academic, Boston.

Hunding, A., Kepes, F., Lancet, D., Minsky, A., Norris, V., Raine, D., Siriam, K., and Root-Bernstein, R. (2006). Compositional complementarity and prebiotic ecology in the origin of life. *BioEssays*, **28**: 399–412.

Huxley, J.P. (1942). *Evolution: the modern synthesis* (reprinted 1963). Allen and Unwin, London.

Idso, S.B. (1986). Industrial age leading to the greening of Earth? *Nature*, **320**: 22.

Impey, C.D.(1995). The search for life in the universe: a humanistic perspective. *Vistas in Astronomy*, **39**: 553–571.

Ingman, M., Kaessmann, H., Pääbo, S., and Gyllensten, U. (2000). Mitochondrial genome variation and the origin of modern humans. *Nature*, **408**: 708–713.

Jacobson, M.Z. (2002). *Atmospheric pollution*. Cambridge University Press, Cambridge.

Jakosky, B.(1998). *The search for life on other planets*. Cambridge University Press, Cambridge.

Jones, B.W.(2004). *Life in the Solar System and beyond*. Springer Praxis, New York.

Kasting, J.F. (2004). When methane made climate. *Scientific American*, **291**: 54–59.

Kasting, J.F. (2006). Earth Sciences: ups and downs of ancient oxygen. *Nature*, **443**: 643–645.

Kasting, J.F. and Siefert, J.L. (2002). Life and the evolution of Earth's atmosphere. *Science*, **296**: 1066–1088.

Kasting, J.F., Whitmire, D.P., and Reynolds, R.T.(1993). Habitable zones around main sequence stars. *Icarus*, **101**: 108–128.

Kateriya, S., Nagel, G,. Bamberg, E., and Hegemann, P. (2004). 'Vision' in single-celled algae. *News in Physiological Science*, **19**: 133–137.

Keeling, C.D., Chin, J.F.S., and Whorf, T.P. (1996). Increased activity of northern vegetation inferred from atmospheric CO_2 measurements. *Nature*, **382**: 146–149.

Keppler, F., Hamilton, J.T.G., Brabs, M., and Rockmann, T. (2006). Methane emissions from terrestrial plants under aerobic conditions. *Nature*, **439**: 187–191.

Kerr, R.A. (2008). Alien planetary system looks a lot like home. *Science*, **319**, 885.

Kiang, N.Y., Segura, A., Tinetti, G., Govindjee, Blankenship, R.E., Cohen, M., Siefert, J., Crisp, D., and Meadows, V.S. (2007). Spectral signatures of photosynthesis. *Astrobiology*, **7**: 252–274.

Kimura, M. (2008). *The neutral theory of molecular evolution*. Cambridge University Press, Cambridge.

Knoll, A.H. (2003). *Life on a young planet*. Princeton University Press, Princeton, NJ.

Kopp, R.E., Kirschvink, J.L., Hilburn, I.A., and Cody, Z.N. (2005). The Paleoproterozoic snowball Earth: a climate disaster triggered by the evolution of oxygenic photosynthesis. *Proceedings of the National Academy of Sciences USA*, **102**: 11131–11136.

Kump, L.R., Pavlov, A., and Arthur, M.A. (2005). Massive release of hydrogen sulfide to the surface ocean and atmosphere during intervals of oceanic anoxia. *Geology*, **33**: 397–400.

Laakso, T., Rantala, J., and Kaasalainen, M. (2006). Gravitational scattering by giant planets. *Astronomy and Astrophysics*, **456**: 373–378.

Lahav, N. (1999). *Biogenesis: theories of life's origin*. Oxford University Press, New York.

Lane, N.(2003). *Oxygen*. Oxford University Press, Oxford.

Lang, K.R.(1997). *Sun, earth and sky*. Springer, Berlin.

Lang, K.R. (1999). The Sun. In: J.K. Beatty, C. Collins Petersen, and A. Chaikin (ed.), *The new Solar System*, pp. 23–38. Cambridge University Press, Cambridge.

Laskar, J., Joutel , F., and Robutel, P.(1993). Stabilization of the Earth's obliquity by the Moon. *Nature*, **361**: 615–617.

Lathe, R.(2004). Fast tidal cycling and the origin of life. *Icarus*, **168**, 18–22.

Lawton, J.H. and May, R.M. (eds) (1995). *Extinction rates*. Oxford University Press, New York.

Leitch, E.M. and Vasisht, G. (1998). Mass extinctions and the Sun's encounters with spiral arms. *New Astronomy*, **3**: 51–56.

Lenton, T.M. (2003). The coupled evolution of life and atmospheric oxygen. In: L.J. Rothschild and A.M. Lister (ed.), *Evolution on planet Earth*, Ch. 3. Academic Press, New York.

Leovy, C.(2001). Weather and climate on Mars. *Nature*, **412**: 245–249.

Leroux, M. (2005). *Global warming, myth or reality. The erring ways of climatology*. Springer, Praxis, London.

Levin-Zaidman, S., Englander, J., Shimoni, E., Sharma, A.K., Minton, K.W., and Minsky, A. (2003). Ringlike structure of the *Deinococcus radiodurans* genome. A key to radioresistance? *Science*, **299**: 254–256.

Levy, O., Appelbaum, L., Legatt, W., Gothilf, Y., Hayward, D.C., Miller, D.J., and Hoegh-Guldberg, O. (2007). Light responsive cryptochromes from a simple multicellular animal, the coral *Acropora millepora. Science*, *318*, 467–470.

Lewontin, R.(2000). *It ain't necessarily so. The dream of the human genome and other illusions.* Review Books, New York.

Lewontin, R.(2000). *The triple helix.* Harvard University Press, Cambridge, MA.

Lineweaver, C.H., Fenner, Y., and Gibson, B.K. (2004). The galactic habitable zone and the age distribution of complex life in the Milky Way. *Science*, **303**: 59–62.

Lomborg, B. (2001). *The skeptical environmentalist.* Cambridge University Press, Cambridge.

Lomborg, B. (2008). *Cool it. The skeptical environmentalist's guide to global warming.* Vintage Books, New York.

Long, S.P., Ainsworth, E.A., Rogers, A., and Ort, D.R. (2004). Rising atmospheric carbon dioxide. Plants FACE the future. *Annual Review of Plant Biol*ogy, **55**: 591–628.

Lovelock, J. (1988). *The ages of Gaia.* Norton and Co., New York.

Lovis, C. *et al.* (2006). An extrasolar planetary system with three Neptune-mass planets. *Nature*, **441**, 305–309.

Lunine, J.I. (1999). *Earth—evolution of a habitable world.* Cambridge University Press, Cambridge.

Lunine, J.I. (2005). *Astrobiology—a multidisciplinary approach.* Pearson, Addison-Wesley, San Francisco.

Macdougall, D. (2004). *Frozen Earth. The once and future story of ice ages.* University of California Press, Berkeley, CA.

McKay, C. P. (2007). To Mars by way of the Moon. *The Planetary Report*, **28**: 12–15.

MacLeod, N. (2003) The causes of Phanerozoic extinctions. In: L.J. Rothschild and A.M. Lister (ed.), *Evolution on Planet Earth*, Ch. 14, Academic Press, New York.

Mancinelli, R.L. (2003). What good is nitrogen: an evolutionary prospective. In: L.J. Rothschild and A.M. Lister (ed.), *Evolution on planet Earth*, Ch. 2, Academic Press, New York.

Margulis, L. and Sagan, D. (1995). *What is life?* University California Press, Berkeley, CA.

Martin, L.D and Meehan, T.J. (2005). Extinction may not be forever. *Naturwissenschaften*, **92**: 1–19.

Martin, W. and Russell, M.J. (2007). On the origin of biochemistry at an alkaline hydrothermal vent. *Philosophical Transactions of the Royal Society B: Biological Sciences*, **362**: 1887–1925.

Mayr, E. (2004). *What makes biology unique?* Cambridge University Press, Cambridge.

Miller, S.L. (1953). Production of amino acids under possible primitive Earth conditions. *Science*, **117**: 528.

Miller, S.L. and Lazcano, A. (2002). Formation of the building blocks of life. In: J.W. Schopf (ed.), *Life's origin*, pp. 78–112, University California Press, Berkeley, CA.

Miller, S.L., and Urey, H.C. (1959). Organic compound synthesis on the primitive Earth. *Science*, **130**: 245.

Monod, J. (1971). *Chance and necessity: an essay on the natural philosophy of modern biology.* Alfred A. Knopf, New York.

Morbidelli, A., Chambers, J., Lunine, J.I., Petit, J.M., Robert, F., Valsecchi, G.B., and Cyr, K.E. (2000). Source regions and timescales for the delivery of water to the Earth. *Meteoritics and Planetary Science*, **35**: 1309–1320.

Morey-Holton, E.R. (2003). The impact of gravity on life. In: L.J. Rothschild and A.M. Lister (ed.), *Evolution on planet Earth*, pp. 143–149. Academic Press, New York.

Morowitz, H. and Sagan, C. (1967). Life in the clouds of Venus. *Nature*, **215**: 1259–1260.

Myneni, R.B., Keeling, C.D., Tucker, C.J., Asrar, G., and Nemani, R.R. (1997). Increased plant growth in the northern high latitudes from 1981 to 1991. *Nature*, **386**: 698–702.

Nagy, F., Fejes, E., Wehmeyer, B., Dallman, G., and Schafer, E. (1993). The circadian oscillator is regulated by a very low fluence response of phytochrome in wheat. *Proceedings of the National Academy of Sciences USA*, **90**: 6290–6294.

Niklas, K.J. (1998). The influence of gravity and wind on land plant evolution. *Review of Paleobotany and Palynology*, **102**: 1–14.

Nisbet, E.G. and Sleep, N.H. (2003). The physical setting for early life. In: L.J. Rothschild and A.M. Lister (ed.), *Evolution on planet Earth*, pp. 3–24. Academic Press, New York.

Norris, V., Hunding, A., Kepes, F., Lancet, D., Minsky, A., Raine, D., Root-Bernstein, R., and Sriram, K. (2007). The first units of life were not simple cells. *Origins of Life and Evolution of Biospheres*, **37**: 429–432.

Nowak, M.A. (2006). *Evolutionary dynamics*. Belknap, Harvard, Cambridge, MA.

Omasa, K., Nouchi, I., and De Kok, L.J. (eds) (2006). *Plant responses to air pollution and global change*. Springer Verlag, Berlin.

Oparin, A.I. (2003). *The origin of life*, Engl. transl. Courier Dover Publications, New York.

Pavlov, A., Hurtgen, M.T., Kasting, J.F., and Arthur, M.A, (2003). Methane-rich Proterozoic atmosphere? *Geology*, **31**: 87–90.

Plaxco, K.W. and Gross, M.(2006). *Astrobiology, a brief introduction*. The Johns Hopkins University Press, Baltimore, MD.

Pohorille, A. (2008). Chemical and biological determinants of habitability. *Astrobiology*, **8**: 331.

Pohorille, A. and Wilson, M.A. (1995). Molecular dynamics studies of simple membrane-water interfaces: structure and functions in the beginning of cellular life. *Origins of Life and Evolution of the Biosphere*, **25**: 21–46.

Prud'homme, B., Gompel, N., and Carroll, S.B. (2007). Emerging principles of regulatory evolution. *Proceedings of the National Academy of Sciences USA*, **104**(Suppl. 1): 8605–8612.

Rachmilevitch, S., Reuveni, J., Pearcy, W.P., and Gale, J. (1999). A high level of atmospheric oxygen, as occurred toward the end of the Cretaceous period, increases leaf diffusion conductance. *The Journal of Experimental Botany*, **50**: 869–872.

Raulin, F. (2005). Astrobiology of Saturn's moon Titan. In: G. Horneck and P. Rettberg (ed.), *Complete course in astrobiology*, Ch. 9. Wiley-VCH, Weinheim.

Raup, D.A. and Sepkoski, J.J. Jr (1984). Periodicity of extinctions in the geologic past. *Proceedings of the National Academy of Sciences USA*, **81**: 801–805.

Rayner, J.M.V.(2003). Gravity, the atmosphere and the evolution of animal locomotion. In: L.J. Rothschild and A.M. Lister (ed.), *Evolution on planet Earth*, pp. 161–179. Academic Press, New York.

Rennie, J. (ed.) (2005). *The water of life*. Scientific American Special Report. Scientific American Publishers, New York.

Reynolds, R.T., McKay, C.P., and Kasting, J.F.(1987). Europa, tidally heated oceans, and habitable zones around giant planets. *Advances in Space Research*, **7**: 125–132.

Richards, J.T., Corey, K.A., Paul, A-L., Ferl, R.J., Wheeler, R.M., and Schuerger, A.C. (2006). Exposure of *Arabidopsis thaliana* to hypobaric environments: implications for low-pressure bioregenerative life support systems for human exploration missions and terraforming on Mars. *Astrobiology*, **6**: 851–866.

Riebe, C.S., Kirchner, J.W., Granger, D.E., and Finkel, R.C. (2001). Strong tectonic and weak climatic control of long term chemical weathering rates. *Geology*, **29**: 511–514.

Rind, D. (2002). The Sun's role in climate variations. *Science*, **296**: 673–677.

Robert, F. (2001). The origin of water on Earth. *Science*, **293**: 1056–1058.

Rohde, R.A. and Muller, R.A. (2005). Cycles in fossil diversity. *Nature*, **434**: 208–210.

Rose, C. and Wright, G. (2004). Inscribed matter as an energy efficient means of communication with an extraterrestrial civilization. *Nature*, **431**: 47–49.

Rothschild, L.J. and Lister, A.M. (eds) (2003). *Evolution on planet Earth*. Academic Press, New York.

Rothschild, L.J. and Mancinelli, R.L. (2001). Life in extreme environments. *Nature*, **409**: 1092–1101.

Roussel, E.G., Cambon Bonavita, M-A., Querellou, J., Cragg, B.A., Webster, G., Prieur, D., and Parkes, R.J. (2008). Extending the sub-sea-floor biosphere. *Science*, **320**: 1046.

Ruiz-Mirazo K., Pereto J., and Moreno A. (2004). A universal definition of life: anatomy and open ended evolution. *Origins of Life and Evolution of the Biosphere*, **34**: 323–346.

Rummel, J.D. and Billings, L. (2004). Issues in planetary protection: policy, protocol and implementation. *Space Policy*, **20**: 49–54.

Rye,R., Kuo, P.H., and Holland, H.D. (1995). Atmospheric carbon dioxide concentrations before 2.2 billion years ago. *Nature*, **378**: 603–605.

Sagan, C. and Chyba, C. (1997). The early faint sun paradox: organic shielding of ultraviolet-labile greenhouse gases. *Science*, **276**: 1217–1221.

Sagan, D. and Margulis, L. (1997). *Microcosmos: four billion years of evolution from our microbial ancestors*. University of California Press, Berkeley, CA.

Scharlemann, J.P.W. and Laurence, W.F. (2008). How green are biofuels? *Science*, **319**: 43–44.

Schmer, M.R., Vogel, K.P., Mitchell, R.B., and Perrin, R.K. (2008). Net energy of cellulosic ethanol from switchgrass. *Proceedings of the National Academy of Sciences USA*, **105**: 464–469.

Schopf, J.W. (2002). *Life's origin*. University of California Press, Berkeley, CA.

Schrödinger, E. (2004). *What is life?* Canto Edition, Cambridge University Press, Cambridge (originally published in 1944).

Schulze-Makuch, D. and Grinspoon, D.H.(2005). Biologically enhanced energy and carbon cycling on Titan. *Astrobiology*, **5**: 560–567.

Schulze-Makuch, D. and Irwin, L.N. (2006). *Life in the universe*. Springer, Berlin.

Schulze-Makuch, D., Grinspoon, D.H., Abbas, O., Irwin, L.N., and Bullock, M.A. (2004). A sulphur based survival strategy for putative phototrophic life in the Venusian atmosphere. *Astrobiology*, **4**: 11–18.

Scott A.C. and Glasspool I.J. (2006). The diversification of Paleozoic fire systems and fluctuations in atmospheric oxygen concentration. *Proceedings of the National Academy of Sciences USA*, **103**: 10861–10865.

Searchinger,T., Heimlich, R., Houghton, R.A., Dong, F., Elobeid, A., Fabiosa, J., Tokgoz, S., Hayes, D., and Yu, T-H. (2008).Use of U.S. croplands for biofuels increases greenhouse gases through emissions from land use change. *Science*, **319**: 1238–1240.

Seckbach, J. (ed.) (2006). *Cellular origin, life in extreme habitats and astrobiology*. Springer Verlag, Berlin.

Seinfeld, J.H., Pandis, H.J., and Spyros, N. (1998). *Atmospheric chemistry and physics—from air pollution to climate change*. John Wiley and Sons, New York.

Shapiro, R. (2007). A simpler origin for life. *Scientific American*, **296**: 24–31.

Shaviv, N.J. (2003a). Towards a solution to the early faint sun paradox: a lower cosmic ray flux for a stronger solar wind. *Journal of Geophysical Research—Space Physics*, **108**: 1437–1445.

Shaviv, N.J. (2003b). The spiral structure of the Milky Way, cosmic rays and ice age epochs on Earth. *New Astronomy*, **8**: 39–77.

Shaviv, N.J. (2005). On the link between cosmic rays and terrestrial climate. *International Journal of Modern Physics*, **20**: 6662–6665.

Shaviv, N.J. and Veizer, J. (2003). Celestial driver of Phanerozoic climate? *GSA Today*, **13**: 4–10.

Shenhav, B., Oz, A., and Lancet, D. (2007). Coevolution of compositional protocells and their environment. *Philosophical Transactions of the Royal Society B: Biological Sciences*, **362**: 1813–1819.

Shindell, D., Rind, D., Balachandran, N., Lean, J., and Lonergan, P. (1999). Solar cycle variability, ozone and climate. *Science*, **284**: 305–308.

Siedow, J.N. (2001). Feeding ten billion people. Three views. *Plant Physiology*, **126**: 20–22.

Simpson, S. (2000). Staying sane in space. *Scientific American*, **282**(3), 61–62.

Singer, S.F. and Avery, D.T. (2007). *Unstoppable global warming, every 1500 years*. Rowman and Littlefield, Lanham, MD.

Smil, V. (2000). *Feeding the world: a challenge to the 21st century*. MIT Press, Cambridge, MA.

Smith, J.M. (1982). *Evolution and the theory of games*. Cambridge University Press, Cambridge.

Sobel, D. (2005). *The planets*. Viking, New York.

Soderberg, A.M. *et al.* (2008). An extremely luminous X-ray outburst at the birth of a supernova. *Nature*, **453**: 469–474.

Somero, G.N., Osmond, C.B., and Bollis, C.L. (eds) (1992). *Water and life*. Springer Verlag, Berlin.

Son, S.-W., Polvani, L.M., Waugh, D.W., Akiyoshi, H., Garcia, R., Kinnison, D., Pawson, S., Rozanov, E., Shepherd, T.G., and Shibata, K. (2008). The impact of stratospheric ozone recovery on the Southern Hemisphere westerly jet. *Science*, **320**: 1486–1489.

Sotin, C. and Prieur, D. (2005). Jupiter's moon Europa: geology and habitability. In: G. Horneck and P. Rettberg (ed.), *Complete course in astrobiology*, Ch. 10. Wiley-VCH, Weinheim

Sterelny, K. (2001). *Dawkins vs. Gould: survival of the fittest*. Icon Books, Cambridge.

Svensmark, H. and Calder, N. (2007). *The chilling stars: the new theory of climate change*. Icon Books, Cambridge.

Swain, M.R., Vasisht, G., and Tinetti, G. (2008). The presence of methane in the atmosphere of an extrasolar planet. *Nature*, **452**: 329–331.

Tartari, A. and Forlani, G. (2008). Osmotic adjustments in a psychrophilic alga, *Xanthonema* sp. (Xanthophyceae). *Environmental and Experimental Botany*, **63**: 342–350.

Taylor, S.R.(1998). *Destiny or chance. Our Solar System and its place in the cosmos*. Cambridge University Press, Cambridge.

Tian, F., Toon, O.B., Pavlov, A.A., and Sterck, H.D. (2005). A hydrogen-rich early Earth atmosphere. *Science*, **308**: 1014–1017.

Todd, P., Anton, H.J., Barlow, P.W., Gerzer, R., Heim, J.-M., Hemmersbach-Krause, R., Slenzka, K., Kordyum, E., Duke, P.J., and Duprat, A.M.(1996). Life and gravity: physiological and morphological responses [Proceedings of the F1.1 Meeting of COSPAR Scientific Commission F held during the Thirtieth COSPAR Scientific Assembly, Hamburg, Germany, 11–21 July 1994]. *Advances in Space Research*, **17**: 1–300.

Tuba, Z. (ed.) (2005). Ecological responses and adaptation of crops to rising atmospheric carbon dioxide. *Journal of Crop Improvement*, **13**: issues 25 and 26.

Tykva, R. and Berg, D. (eds) (2004). *Man made and natural radioactivity in environmental pollution and radiochronology*. Kluwer Academic Press, Dordrecht.

Valley, J.W., Peck, W.H., King, E.M., and Wilde, S.A. (2002). A cool early Earth. *Geology*, **30**: 351–354.

Van Allen, J.A. and Bagenal, F. (1999). Planetary magnetospheres and the interplanetary medium. Beatty, J.K., Collins Petersen, C., and Chaikin, A.(eds) (1999) *The new Solar System*, pp. 39–58. Cambridge University Press, Cambridge.

Veizer, J. (2005). The celestial climate driver. A perspective from four billion years of the carbon cycle. *Geoscience-Canada*. **32**: 13–28.

Volk, T. (2998). *Gaia's body*. Springer Verlag, Berlin.

Walker, G. (2003). *Snowball Earth*. Crown Publishers, New York.

Walker J.C.G. (1994) Global geochemical cycles of carbon. In: N.E. Tolbert and J. Preiss (eds), *Regulation of atmospheric CO_2 and O_2 by photosynthetic carbon metabolism*, pp. 75–79. Oxford University Press, Oxford.

Ward, P.D. and Brownlee, D.(2000). *Rare Earth: why complex life is uncommon in the universe*. Copernicus Books, Springer Verlag, New York.

Warmflash, D. and Weiss, B. (2005). Did life come from another world? *Scientific American*, **295**: 40–47.

Wassersug, R.J. (1999). Life without gravity. *Nature*, **401**: 758.

Watson, A., Lovelock, J.E., and Margulis, L. (1978). Methanogenesis, fires and the regulation of atmospheric oxygen. *Biosystems*, **10**: 293–298.

Webb, S. (2002). *Where is everybody? Fifty solutions to the Fermi paradox*. Copernicus Books, Springer Verlag, New York.

Weissman, P.S., McFadden, L-A., and Johnson, T.V. (eds) (1999). *Encyclopedia of the Solar System*. Academic Press, New York.

Wetherill, G.W. (1995). How special is Jupiter? *Nature*, **373**: 470.

White. R.J. and Averner, M. (2001). Humans in space. *Nature*, **409**: 1115–1118.

Wilson, E.O. (2002). *The future of life*. Alfred A. Knopf, New York.

Wittwer, S.H. (1995). *Food, climate and carbon dioxide*. CRC Press, Boca Raton, FL.

Wolpert D.H. and Macready W. (2004). Self-dissimilarity as a high dimensional complexity measure. In: Y. Bar-Yam (ed.), *Proceedings of the International Conference on Complex Systems*. Perseus Books, New York. (Available at: http://ti.arc.nasa.gov/m/pub/889h/0889%20(Wolpert).pdf)

Wong, C.E., Li, Y., Whitty, B.R., Díaz-Camino, C., Akhter, S.R., Brandle, J.E., Golding, G.B., Weretilnyk, E.A., Moffatt, B.A., and Griffith, M. (2005). Expressed sequence tags from the Yukon ecotype of *Thellungiella* reveal that gene expression in response to cold, drought and salinity shows little overlap. *Plant Molecular Biology*, **58**: 561–574.

Xiong, J. and Bauer, C.E. (2002). Complex evolution of photosynthesis. *Annual Review of Plant Biology*, **53**: 503–521.

Zhang, D.D., Zhang, J., Lee, H.F., and He, Y.-Q. (2007) Climate change and war frequency in eastern China over the last millennium. *Human Ecology*, **35**: 403–414.

Zierenberg, R.A., Adams, M.W.W., and Arp, A.J. (2000). Life in extreme environments. Hydrothermal vents. *Proceedings of the National Academy of Sciences USA*, **97**: 12961–12962.

Zimmer, C. (1999). Complex systems: life after chaos. *Science*, **284**: 83–86.

Zimmer, C. (2008). What is a species? *Scientific American*, **298**: 48–55.

Astrobiology and related journals and newsletters

Advances in Space Research, Elsevier, Amsterdam

Astrobiology. MaryAnn Liebert Publications, New Rochelle, NY

International Journal of Astrobiology. Cambridge University Press, Cambridge

Origin of Life and Evolution of Biospheres. Springer Verlag, Berlin
The Planetary Report. The Planetary Society, Pasadena, CA (members' newsletter of the Planetary
   Society, see website below)

Internet portals

Note: none are specifically oriented to the astrobiology of *Earth.*

- The Internet Encyclopedia of Science maintained by astronomer and author David Darling: http://www.daviddarling.info/encyclopedia/E/ETEnotes.html
- For a brief updated commentary on Schrödinger's classic book *What is life?* see: http://www.whatislife.com/about.html
- Life in the universe—current readings and references: http://www.mainsgate.com/spacebio/modules/lu_read.html
- For the US National Research Council report *The limits of organic life in planetary systems*: http://www.nap.edu/catalog.php?record_id=11919
- NASA Astrobiology Institute: http://astrobiology.nasa.gov/nai/
- The main NASA astrobiology newsletter: http://www.astrobio.net/news/
- Astrobiology workshops: http://astrobiology.arc.nasa.gov/workshops/
- Astrobiology roadmap: http://astrobiology.arc.nasa.gov/roadmap/
- SETI (Search for Extraterrestrial Intelligence) Institute: http://www.seti.org/
- The Astrobiology Web: http://www.astrobiology.com/
- Eric Weissteins' Treasure Trove of Science: a collection of definitions and articles. See 'Seasons' for animated explanation of seasons and Earth's tilt: http://www.treasure-troves.com
- Many short but very good articles on astrobiology can be found in a special edition of the popular astronomy journal *Ad Astra*. This edition is available at: http://www.astrobiology.com/adastra/astrobiology.101.html
- A most comprehensive and up-to-date review of what is known of Saturn's moon Titan can be found in Wikipedia: http://en.wikipedia.org/wiki/Titan_moon
- For the Jovian moon Europa: http://en.wikipedia.org/wiki/Europa_moon
- For Saturn's Enceladus: http://en.wikipedia.org/wiki/Enceladus_(moon)
- For a discussion of methane findings on Titan: http://www.nasa.gov/mission_pages/cassini/media/20020727b.html (Cassini homepage http://www.nasa.gov/mission_pages/cassini/main/index.html gives links to other articlesof interest and recent findings)
- The Planetary Society website: http://planetary.org/home/
- For the latest Planetary Society news see: http://planetary.org/news/index.html
- For the seminal Sagan/Mayr debate on the chances of life in the cosmos: http://www.planetary.org/explore/topics/search_for_life/seti/seti_debate.html
- European Space Agency, human spaceflight and exploration news: http://www.esa.int/esaHS/index.html
- PLOS Genetics—an entry into recent research on the origins of humans: http://genetics.plosjournals.org/perlserv/?request=get-document&doi=10.1371/journal.pgen.0030104
- The UN has a group working on modern world populations: http://www.un.org/esa/population/unpop.htm
- Some useful portals to CELSS research: http://www.permanent.com/s-bios3.htm http://www.permanent.com/s-ce-nhu.htm   http://en.wikipedia.org/wiki/Controlled_Ecological_Life_Support_System

- Three useful sites for further information on plate tectonics are: http://www.platetectonics. com/ http://en.wikipedia.org/wiki/Plate_tectonics http://www.ucmp.berkeley.edu/geology/ techist.html
- For tectonics and the carbon cycle: http://dilu.bol.ucla.edu/
- A short discussion of the putative importance of the formation of the Moon to Earth tectonics and sea/land separation can be found at: http://www.spacedaily.com/news/life-01x1.html
- International Panel on Climate Change (IPCC): http://www.ipcc.ch/
- Much information relating to the argument on CO_2 versus extraterrestrial contributions to global heating can be found in: http://www.sciencebits.com/CO2orSolar
- For information on 'the shadow biosphere' see: http://www.astrobio.net/news/article2161. html
- A comprehensive and up to date (as of 2008) article on the origin of life: http://en.wikipedia. org/wiki/Origin_of_life
- For much useful material on Evo-Devo see: http://www.seanbcarroll.com/

Index

Appendix, figures, notes and tables are indexed in bold.